Photographic and Descriptive Musculoskeletal Atlas of *Gorilla*

With notes on the attachments, variations, innervation, synonymy and weight of the muscles

Photographic and Descriptive Musculoskeletal Atlas of *Gorilla*

With notes on the attachments, variations, innervation, synonymy and weight of the muscles

- Rui Diogo
- Juan F. Pastor
- Eva M. Ferrero
- Mercedes Barbosa

- Josep M. Potau
- Félix J. de Paz
- Gaëlle Bello
- Bernard A. Wood

With a Foreword by Russell H. Tuttle

CRC Press
Taylor & Francis Group
Boca Raton London New York

CRC Press is an imprint of the
Taylor & Francis Group, an **informa** business

A SCIENCE PUBLISHERS BOOK

CRC Press
Taylor & Francis Group
6000 Broken Sound Parkway NW, Suite 300
Boca Raton, FL 33487-2742

First issued in paperback 2017

© 2011 by Taylor & Francis Group, LLC
CRC Press is an imprint of Taylor & Francis Group, an Informa business

ISBN 13: 978-1-138-11323-7 (pbk)
ISBN 13: 978-1-57808-694-8 (hbk)

Library of Congress Cataloging-in-Publication Data

Photographic and descriptive musculoskeletal atlas of gorilla : with
notes on the attachments, variations, innervation, synonymy, and weight
of the muscles / [edited by] Rui Diogo ... [et al.]. ; with a foreword
by Russell H. Tuttle -- 1st ed.
     p. cm.
  Includes bibliographical references and index.
  ISBN 978-1-57808-694-8 (hardback)
1. Gorilla--Atlases. 2. Gorilla--Muscles--Atlases. 3.
Musculoskeletal system--Atlases. I. Diogo, Rui.
  QL737.P96P49 2010
  573.7'19884--dc22

                                                                2010033460

Foreword

Diogo and colleagues have prepared a superb subcutaneous portrait of one of the most magnificent species of Mammalia on Earth: *Gorilla gorilla*, vernacularly known as western gorillas.

Medical students in gross anatomy are often befuddled by the fact that the structures that are so clearly illustrated in classic human anatomy textbooks and manuals are so much clearer than those of their cadavers. The reason is that the best illustrations of human structure were prepared with deceased individuals that had not been embalmed. The same is true in the gorilla atlas of Diogo et al.

As genomics overwhelm classic fields of comparative, functional and evolutionary biology, I fear that the insights that come from the art of meticulous dissection so well illustrated in this atlas will lapse, and that premedical students and newer generations of evolutionary biologists will be even less appreciative of whole organisms with myriad functions and potential based not only on their genomic heritage, but perhaps more importantly on the anatomical bases of their functional capacities and versatility, the latter of which might enable them to survive in ever changing habitats.

Less practically, but equally interesting is the potential of the atlas to serve modelers of primate, and particularly hominoid phylogenies. Evolutionary clocks are set by paleontological evidence, which is purely morphological. The ability to recognize hints of muscle markings on fragmentary fossils depends on detailed knowledge of the muscles of extant species. For example, *Pongo* spp. bilaterally lack the anterior bellies of the digastric muscles that in all other extant Hominoidea attach distally to fossae (concave depressions) on the border near the mandibular symphysis (Fig. 7). Although the premolar and molar teeth of 7–9 Ma *Khoratpithecus piriyai* from Thailand closely resemble those of 8–9 Ma Chinese specimens of *Lufengpithecus lufengensis* and harbinger *Pongo* spp., the absence of facets for the attachment of the anterior bellies of the digastric muscles on RIN 765 is a feature that *Khoratpithecus piriyai* shares uniquely with *Pongo* (Chaimanee et al. 2004). In this instance, presence or absence of digastric impressions on fossil mandibles should be very helpful to modelers of the evolution of *Pongo* and bound speculations on when and where they diverged from the lineages of other apes and humans.

RUSSELL H. TUTTLE

Professor
Department of Anthropology
Committee on Evolutionary Biology
Morris Fishbein Center for the History of Biology and Medicine
and the College
The University of Chicago.
May 31, 2010

Reference cited

Chaimanee, Y., Suteethorn, V., Jintasakul, P., Vidthayanon, C., Marandat, B., and Jaeger, J.-J. 2004. A new orang-utan relative from the Late Miocene of Thailand. *Nature* 427: 439–441.

Acknowledgements

BW gratefully acknowledges support and funding from GWU's VP for Academic Affairs Don Lehman and from GWU's Signature Program Fund. RD was generously supported by a GW Presidential Merit Fellowship.

Contents

CHAPTER 1

Introduction and Aims

Gorillas, apart from common chimpanzees and bonobos, are our closest living relatives. Therefore, it is crucial to understand their biology, including their anatomy, to help provide a context for our own evolutionary history. However, a review of the literature (see Annex I) clearly indicates that the anatomy of gorillas, and in particular their musculature, has been relatively neglected compared to the information available for chimpanzees. Raven (1950) is usually cited as the main source of information, but it should be emphasized that the descriptions therein were mainly based on the dissection of a single specimen, and we now know that the information it provides is far from comprehensive. For example, about half of the head and neck muscles that are usually present in gorillas (see Chapter 3) were not described by Raven (e.g., the levator veli palatini, tensor veli palatini, tensor tympani, stapedius, stylopharyngeus, constrictor pharyngis medius, constrictor pharyngis inferior, cricothyroideus, constrictor pharyngis superior, palatopharyngeus, thyroarytenoideus, cricoarytenoideus lateralis, arytenoideus obliquus, arytenoideus transversus, cricoarytenoideus posterior, genioglossus, hyoglossus, styloglossus, palatoglossus, and the intrinsic facial muscles of the ear and the intrinsic tongue muscles).

This lack of published and easily accessible information about the musculature of gorillas (which is in large part related to the difficulty of obtaining specimens for careful anatomical dissection) has implications for our understanding of the functional morphology and evolution of primates. How can one comprehensively discuss the evolution of the modern human pharyngeal region, or the tongue, or of functions such as speech when there is virtually no information about the pharyngeal, laryngeal and tongue muscles of closely-related primates such as the gorilla?

Another implication concerns the use of myological data to generate phylogenetic hypotheses among hominoids, and among primates as a whole. The results of earlier analyses have suggested that myological characters are particularly effective for recovering phylogenetic relationships within hominoids (Gibbs 1999; Gibbs et al. 2000, 2002) and other vertebrate groups (Diogo 2004a,b, 2007). By using myological data as the basis of characters in cladistic analyses, Diogo (2004a,b, 2007) has shown that myology-based characters are capable of reconstructing the phylogenetic relationships (particularly of higher taxa such as families and orders) that match the pattern of relationships which are well-supported by other lines of evidence (including, e.g., osteological and/or molecular evidence). One of the reasons why myological characters may be particularly effective at recovering the phylogeny of higher taxa may be related to the results of the experiments that used rhombomere quail-to-chick grafts to investigate the influence of hindbrain segmentation on craniofacial patterning (Köntges and Lumsden 1996). This experimental study showed that each rhombomeric population remains coherent throughout the ontogeny, with rhombomere-specific matching of muscles and their attachment sites for all branchial and tongue muscles. If a similar system operates elsewhere in the body, it would help explain how muscle gross morphology is conserved, whereas the shapes of the skeletal elements to which the muscles are attached are susceptible to changes that might contrive to obscure phylogeny (for reviews, see Diogo 2007 and Diogo and Abdala 2010).

The first photographic and descriptive musculoskeletal atlas of *Gorilla* is provided here. There are many ways in which this atlas differs from that of Raven (1950), as well as from similar works that have been published for other primates such as baboons and chimpanzees (e.g., Swindler and Wood 1973).

First, we were able to dissect and take high-quality photographs of four fresh adult gorilla specimens (see Methodology and Material below). We were also able to dissect, photograph, and measure the weight of the majority of muscles of the female specimen VU GG1, and this specimen provided most of the photographs and all the weights of the muscles in this atlas. In instances where there were differences between the myological data obtained from this specimen and the data obtained from the other three specimens (e.g., the presence/absence of a muscle or a muscle bundle, their attachments and/or their innervation) we provide detailed comparative notes and photographs, to document these differences.

Second, the atlas includes not only the results of our own dissections of these four specimens, but also the results of an extensive review of the literature, including reports published more than 150 years ago by authors such as Vrolik (1841), Duvernoy (1855–1857), Huxley (1864) and Owen (1868). Thus, by highlighting studies that differ from our own observations the atlas provides a comprehensive review of muscular variants among individual gorillas. It also includes an extensive list of the synonyms used in the literature to label the muscles of these primates.

Third, data previously obtained from our dissections of numerous primates and other mammalian and non-mammalian vertebrates (e.g., Diogo et al. 2008, 2009a,b; Diogo and Abdala 2010) were used to test hypotheses about the homologies between the muscles of gorillas, modern humans, and other taxa.

It is hoped that this atlas will be of interest to students, teachers and researchers working in primatology, comparative anatomy, functional morphology, zoology, or physical anthropology, as well as to clinicians and researchers who are interested in understanding the origin, evolution, homology and variations of the musculoskeletal system of modern humans.

CHAPTER 2

Methodology and Material

We dissected four fresh adult specimens of *Gorilla gorilla*. They were made available by the Canadian Museum of Nature (specimen CMS GG1, male) and Valladolid University (specimen VU GG1, female; specimen VU GG2, female; specimen VU GG3, male). Photographs were taken of the musculoskeletal system of all four specimens, but, as explained above, most of the photographs and all the masses of the muscles listed in this atlas are from the female specimen VU GG1. The reason being that this 98 kg. specimen was in particularly good condition, and we were able to carefully dissect, photograph, and weigh most of its muscles. The photographs of the osteological structures shown in this atlas are from the male specimen VU GG3.

For each muscle, when the information is available, we provide: 1) the mass in grams of the muscle in the VU GG1 specimen (when the muscle is paired, e.g., stylohyoideus, the mass given is that of the muscle of one side of the body; when the muscle is unpaired, e.g., arytenoideus transversus, the mass given is that of half of the muscle, i.e., also from a single side of the body). It is not possible to accurately measure the mass of all the muscles, but when it can be measured the value we obtained is given in parentheses immediately following the name of the muscle; 2) the most common attachments and innervation of the muscle within gorillas, based on our dissections of the four specimens and on the results of our literature review; 3) comparative notes, for instance where there are differences (e.g., regarding the presence/absence of the muscle, or of its bundles, its attachments, and/or its innervation) between the configuration usually found in gorillas and that found in a specimen dissected by us (in these cases we often provide photographs to illustrate the differences) or by others; 4) a list of the synonyms other authors have used for that muscle.

Apart from the four gorilla specimens mentioned above, we have dissected numerous specimens from most vertebrate groups, including bony fish, amphibians, reptiles, monotremes, rodents, colugos, tree-shrews, and numerous primates, including humans (a complete list of these specimens and the terminology used to describe them is given by Diogo and Abdala 2010). This proved to be crucial for examining the homologies between the muscular structures of gorillas, modern humans, and other primate and non-primate vertebrates, and, thus, for using the nomenclature proposed by Diogo et al. (2008, 2009a,b) and Diogo and Abdala (2010). This nomenclature is mainly based on that employed in modern human anatomy (e.g., Terminologia Anatomica 1998), but also takes into account the names used by researchers working with non-human mammals (e.g., Saban 1968; Jouffroy 1971), by integrating what is known about the origin, evolution and homologies of the muscles reported here. In the illustrations provided in this book, Latin names are normally used to label the musculoskeletal structures that are shown. In order to avoid redundancy, when these names are similar to the English names (e.g., processus mastoideus and mastoid process), the English names are not provided. However, in those cases in which they are substantially different (i.e., the readers who are not familiar with Latin may not be able to deduce the equivalent English name: e.g., incisura mandibulae and mandibular notch), both the Latin names and the English names are provided.

When the position, attachments and orientation of the muscles are described and the terms anterior, posterior, dorsal and ventral are used, it is done in the sense that the terms are applied to pronograde tetrapods (e.g., the sternohyoideus mainly runs from the sternum, posteriorly, to the hyoid bone, anteriorly, and passes ventrally to the larynx, which is, in turn, ventral to the esophagus; the flexors of the forearm are mainly situated on the ventral side of the forearm). However, the nomenclature used in Terminologia Anatomica (1998) was defined on the basis of an upright, or orthograde, posture, and, even if most primates are not upright bipeds, all the osteological names, and most of the myological ones, used by other authors (and used by us here) to label the structures of non-human primates, including gorillas, use this nomenclature. This could seem confusing in certain cases, but we do feel that this is the correct way to describe the topology of the musculoskeletal structures of non-human primates, because in the vast majority of primates the 'superior angle of the scapula' is actually anterior, and not superior, to the 'inferior angle of the scapula', and the 'cricoarytenoideus posterior' actually lies on the dorsal, and not on the posterior, surface of the larynx. Moreover, we think that, by keeping in mind that the names (both in Latin and in English) of all of the osteological and of most the myological structures mentioned in this atlas refer to an orthograde posture while the descriptions provided here regarding the topology of these structures refer to a pronograde posture, the reader should have no major problems in understanding the information provided in this book.

The muscles listed below are those that are usually present in adult gorillas; muscles that are only occasionally present in adult gorillas (e.g., palmaris longus, epitrochleoanconeus, contrahentes digitorum: see Chapter 4) are not listed separately, but are discussed in other parts of the atlas. In our descriptions, we follow Edgeworth (1935) and Diogo and Abdala (2010) and divide the head and neck muscles into five main subgroups: 1) mandibular muscles, which are generally innervated by the fifth cranial nerve (CN5) and include the masticatory muscles, among others; 2) hyoid muscles, which are usually innervated by CN7 and include the facial muscles; 3) branchial muscles, which are usually innervated by CNC9 and CN10, and include most laryngeal and pharyngeal muscles; 4) hypobranchial muscles, which include all the infrahyoid and tongue muscles, and the geniohyoideus. According to Edgeworth (1935) the hypobranchial muscles develop primarily from the anterior myotomes of the body, and then migrate into the head; although they retain a main innervation from spinal nerves, they may also be innervated by CN11 and CN12, but they usually do not receive any branches from CN5, CN6, CN7, CN8, CN9 and CN10; 5) extra-ocular muscles, which are usually innervated by nerves CN3, CN4 and/or CN6 in vertebrates. The head, neck, pectoral and upper limb muscles are listed following the order used by Diogo et al. (2008, 2009a) and Diogo and Abdala (2010), while the pelvic and lower limb muscles, as well as the other muscles of the body, are listed following the order used by Gibbs (1999).

CHAPTER 3

Head and Neck Musculature

3.1 Mandibular musculature

Mylohyoideus (7.24 g; Figs. 7, 8)
- Usual attachments: From the mylohyoid lines of the mandible to the hyoid bone, posteriorly, and to the ventral midline, anteriorly.
- Usual innervation: Mylohyoid nerve of mandibular division of CN5 (Raven 1950).
- Notes: There is usually no distinct median raphe of the mylohyoideus in gorillas, according to Saban (1968), Göllner (1982), Gibbs (1999), and to our dissections.

Digastricus anterior (11.05 g; Figs. 6, 7)
- Usual attachments: From the intermediate tendon of the digastric, and sometimes also from the hyoid bone, to mandible.
- Usual innervation: Mylohyoid nerve of mandibular division of CN5 (Sommer 1907, Raven 1950, Gibbs 1999).
- Notes: In the fetal specimen of Deniker (1885) the digastricus anterior does not contact its counterpart for most of its length, and, probably is based on this work, Gibbs et al. (2002) suggested that this is the usual condition for gorillas. However, the muscle appears to contact its counterpart for most of its length in the gorillas illustrated by Bischoff (1880) and by Hosokawa and Kamiya (1961-1962), and this latter configuration was also present in our VU GG1 specimen (see Fig. 7).

Tensor tympani
- Usual attachments: From the auditory tube and adjacent regions of neurocranium to the manubrium of the malleus.
- Usual innervation: Data not available.
- Notes: According to Maier (2008), in gorillas the chorda tympani usually passes above the tensor tympani (epitensoric condition).

Tensor veli palatini (1.54 g; Figs. 11, 12)
- Usual attachments: From the eustachian process and the scaphoid fossa to the pterygoid hamulus and soft palate.
- Usual innervation: Data not available.
- Notes: As explained by Aiello and Dean (1990), in modern human infants and great apes, including gorillas, the palate lies much closer to the roof of the nasopharynx than it usually does in adult modern humans, so in the great apes the levator veli palatini and tensor veli palatini do not run so markedly downwards to reach the palate as they do in the latter.

Masseter (106.41 g; Figs. 4, 5, 6, 7, 8, 9)
- Usual attachments: Pars superficialis from the zygomatic arch to the lower three-quarters of the ramus of the mandible; pars profunda from the inferior border of the zygoma, zygomatic arch and temporalis fascia to the superior quarter of the ramus and part of the coronoid process of the mandible, some of its fibers are inserted on the tendon of the temporalis.
- Usual innervation: Masseteric nerve of mandibular division of CN5 (Raven 1950, Gibbs 1999).
- Notes: According to Gibbs (1999), contrary to *Homo* and *Pan*, in gorillas, as well as in orangutans, there is no strong aponeurosis or fascial separation between the superficial and deep portions of the masseter. Göllner (1982) dissected two infant gorillas and found a '**zygomatico-mandibularis**' fused with the deep fibers of the masseter and the superficial fibers of the temporalis; however, there was no evidence of a distinct 'zygomatico-mandibularis' in our VU GG1 specimen (see Figs. 5, 6, 9).

Temporalis (217.74 g; Figs. 5, 6, 7)
- Usual attachments: From the whole fossa temporalis and temporalis fascia to the medial and anterior borders of the coronoid process and anterior margin of the ramus of the mandible.
- Usual innervation: Masseteric nerve of the mandibular division of CN5 (Raven 1950, Gibbs 1999).
- Notes: Göllner (1982) described a pars suprazygomatica in neonatal gorillas, but suggested that this structure is usually not differentiated in adults; however, on both sides of our VU GG1 specimen, there was a distinct pars suprazygomatica (see Figs. 5, 7). In his Plate 13, Raven (1950) shows a pars superficialis and a pars profunda of the temporalis, but he does not describe these two bundles in the text. We did not find a distinct pars profunda and a distinct pars superficialis in our VU GG1 specimen (see Figs. 5, 6, 7).

Pterygoideus lateralis (10.36 g; Fig. 10)
- Usual attachments: From the sphenoid fissure, lateral pterygoid plate and adjacent regions of the neurocranium to the capsule of the temporomandibular joint and neck of the mandibular condyle.

- Usual innervation: Branches of mandibular division of CN5 (Raven 1950, Gibbs 1999).
- Notes: Raven (1950) and Göllner (1982) do not refer to a differentiation of the pterygoideus lateralis into a caput inferius and a caput superius, but these two bundles were clearly present in our VU GG1 specimen (see Fig. 10).
- Synonymy: Pterygoideus externus (Raven 1950).

Pterygoideus medialis (44.76 g; Fig. 10)
- Usual attachments: From the pterygoid fossa and the lateral pterygoid plate to the medial and ventral margins of the base of the ramus of the mandible.
- Usual innervation: Branches of mandibular division of CN5.
- Synonymy: Pterygoideus internus (Raven 1950).

3.2 Hyoid musculature

Stylohyoideus (1.87 g; Figs. 6, 7)
- Usual attachments: From the styloid process to the hyoid bone.
- Usual innervation: Branches of CN7 (Sommer 1907, Raven 1950, Gibbs 1999).
- Notes: In gorillas, the stylohyoideus is usually pierced by the intermediate tendon of the digastric, as described by Deniker (1885), Raven (1950) and Gibbs (1999), and also found in our VU GG1 specimen (see Fig. 7). Dean (1984) describes a gorilla in which the stylohyoideus does not seem to originate from the styloid process, but instead from the temporal bone immediately adjacent to this process.

Digastricus posterior (5.87 g; Figs. 6, 7)
- Usual attachments: From the mastoid portion of the temporal bone and lateral border of the occipital bone to the intermediate tendon, which is held close to the hyoid bone by a ligamentous sling/fibrous loop similar to one usually found in modern humans (see, e.g., Plate 59 of Netter 2006).
- Usual innervation: Branches of CN7 (Sommer 1907, Raven 1950, Gibbs 1999).

Stapedius
- Usual attachments: Probably inserts onto the stapes, but the information provided in the literature is very scarce.
- Usual innervation: Data not available.

Platysma myoides (58.06 g; Figs. 1, 2, 3, 4, 15, 18, 53, 54)
- Usual attachments: From the pectoral fascia and from the fascia over the deltoideus and the side and back of the neck (but not from the nuchal region, see below), as well as from the acromial and clavicular regions, to the mandible, depressor anguli oris, depressor labii inferioris, modiolar region and adjacent regions of the mouth.

- Usual innervation: Branches of CN7.
- Notes: Deniker (1885) described a **platysma cervicale** in the gorilla fetus dissected by him (see, e.g., his plate XXVI), and Bischoff (1880), Chudzinski (1885) and Raven (1950) suggested that they also found this muscle in juvenile and adult gorillas (corresponding for instance to the 'nuchal and deep portions of the platysma' described by Raven 1950). However, in works that focused on facial musculature, such as Ruge (1887b) and Huber (1930, 1931), it was suggested that in gorillas the platysma cervicale is usually reduced in size, or even absent (see, e.g., the illustrations of these authors). Our detailed observations of the VU GG1 specimen corroborated this latter suggestion. That is, at first sight the platysma cervicale seems to be present, passing just inferiorly to the external ear. However, more careful observation reveals that this corresponds to the platysma myoides of modern humans because, posteriorly to the external ear, the muscle is oriented inferiorly (see Fig. 1), as is usually the case in modern humans and as shown in the gorillas illustrated in Fig. 50 of Hartmann (1886) and Fig. 36 of Huber (1930). That is, the platysma myoides runs inferiorly to partially cover the pectoralis major, the deltoideus, and the acromial region, but it does not extend posteriorly to reach the nuchal region (see Fig. 1). It should be noted that the **sphincter colli superficialis** and **sphincter colli profundus** are not present as distinct muscles in gorillas.

Occipitalis
- Usual attachments: From the occipital region to the galea aponeurotica and the region of the external ear.
- Usual innervation: Branches of CN7.
- Notes: Ruge (1887b) reported a 'nuchal/neck' portion of the occipitalis in *Gorilla*, which possibly corresponds to the '**cervico-auriculo-occipitalis**' described by Lightoller (1928, 1934, 1939) in some other primate and non-primate mammals. However, the structure reported by Ruge (1887b) was very small and did not reach the external ear, i.e., it appeared to be a 'vestigial' structure (see his Fig. 1). Seiler (1976) described a 'pars profunda' and a 'pars superficialis' of the occipitalis in gorillas, and it is possible that the latter bundle also corresponds to a 'vestigial cervico-auriculo-occipitalis', because this bundle is somewhat superficial/lateral to the main body of the occipitalis, but also does not reach the external ear (see his Fig. 148).
- Synonymy: Auriculo-occipitalis (Ruge 1887b, Raven 1950); occipito-auricularis (Edgeworth 1935).

Auricularis posterior
- Usual attachments: From the occipital region to the posterior portion of the external ear.
- Usual innervation: Branches of CN7.
- Synonymy: Retrahens aurem (Macalister 1873, Hartmann 1886, Ruge 1887b).

Intrinsic facial muscles of ear
- Usual attachments: See notes below.
- Usual innervation: Branches of CN7.
- Notes: The intrinsic facial muscles of the external ears of gorillas have hardly been described in the literature, and were difficult to analyze in our specimens, but Seiler (1976) did examine these muscles in some detail. According to him, the muscles of gorillas are in general somewhat similar to those of modern humans, but there are some differences. He stated that in gorillas the **depressor helicis** and the **tragicus** are inconstantly present. The **helicis minor, antitragicus, obliquus auriculae** and **transversus auriculae** are usually present, as is usually the case in modern humans. He suggested that the **incisurae terminalis** ('**incisurae Santorini**') and the '**intercartilagineus**' are usually absent in gorillas, but that the **pyramidalis auriculae** ('**trago-helicinus**') is usually present in these primates. Hartmann (1886) has also shown a helicis minor in a gorilla specimen (see his Fig. 50). Ruge (1887a,b) reported a '**musculus auriculae proprius posterior**' in *Gorilla* and in *Pan*, which according to him is mainly continuous and not divided into a transversus auriculae and an obliquus auriculae; however, as stated above, Seiler (1976) did describe, and show, a distinct obliquus auriculae and a distinct transversus auriculae in gorillas (see his Fig. 148). The **mandibulo-auricularis** is not present as a distinct, fleshy muscle in gorillas, possibly corresponding to part or the totality of the stylomandibular ligament, as is usually the case in modern humans. Detailed studies of the external ear and its muscles in gorillas are clearly needed.

Risorius (0.49 g; Figs. 2, 3)
- Usual attachments: From the platysma myoides or the region just superior to it, to the angle of the mouth.
- Usual innervation: Branches of CN7 (Raven 1950).
- Notes: Most of the authors who have studied the facial muscles of gorillas described and/or illustrated a risorius (see, e.g., Fig. 50 of Hartmann 1886, Fig. 1 of Ruge 1887b, Fig. 1 of Chudzinski 1885, Plate 9 of Raven 1950 and Fig. 146 of Seiler 1976; see also descriptions of Ehlers 1881, Deniker 1885 and Huber 1930, 1931, although the latter author stated that this muscle was not a 'true' risorius). On both sides of our VU GG1 specimen, there was a structure that was differentiated from the platysma myoides and that seemed to correspond to the muscle risorius which is usually found in modern humans (see Figs. 2, 3), because: 1) it was superficial to the platysma myoides and 2) anteriorly, it was somewhat curved in an antero-inferior direction, while the fibers of the platysma myoides were mainly directed anteriorly. This structure was in general similar to the structure shown in Fig. 1 of Ruge (1887b), running from the superior portion of the platysma myoides to the angle of the mouth. According to the review of Seiler 1976, the risorius is inconstantly present in gorillas.

Zygomaticus major (1.55 g; Figs. 2, 3)

- Usual attachments: From the zygomatic arch/bone (relatively far from the external ear and near the eye, as is usually the case in modern humans: see, e.g., Plate 26 of Netter 2006) to the corner of the mouth, being mainly superior to the platysma myoides and superficial (lateral) to the levator anguli oris facialis.
- Usual innervation: Branches of CN7.
- Notes: Huber (1931) stated that the zygomaticus major of gorillas is not homologous to that of modern humans; however, this statement was contradicted by most of the other authors, as well as by our dissections (see, e.g., Diogo et al. 2009b). According to Deniker (1885) and Raven (1950), in gorillas the zygomaticus major may be divided into two bundles, but we did not find two distinct bundles in our specimen VU GG1 (see Figs. 2, 3).
- Synonymy: Zygomaticus (Ruge 1887b, Sullivan and Osgood 1925, Raven 1950); part of zygomaticus (Edgeworth 1935); zygomaticus inferior (Seiler 1976).

Zygomaticus minor (0.65 g; Figs. 2, 3)

- Usual attachments: From the zygomatic bone and orbicularis oculi (relatively far from the external ear and near the eye, as is usually the case in modern humans: see, e.g., Plate 26 of Netter 2006) to the corner of the mouth and upper lip, being mainly superficial (lateral) to the levator anguli oris facialis.
- Usual innervation: Branches of CN7.
- Synonymy: Orbicularis labialis or orbito-labialis (Ruge 1887b, Sullivan and Osgood 1925); part of the zygomaticus Edgeworth 1935); caput zygomaticum of quadratus labii superioris (Raven 1950); zygomaticus superior (Seiler 1976).

Frontalis (8.46 g; Figs. 2, 3)

- Usual attachments: From galea aponeurotica to the skin of the eyebrow and nose.
- Usual innervation: Branches of CN7.

Auriculo-orbitalis (2.71 g; Figs. 2, 3)

- Usual attachments: From the anterior portion of the external ear to region of frontalis.
- Usual innervation: Posterior auricular nerve, of CN7 (Raven 1950).
- Notes: In Terminologia Anatomica (1998), the **temporoparietalis** is considered to be a muscle that is usually present in modern humans (originating mainly from the lateral part of the galea aponeurotica, and passing inferiorly to insert onto the cartilage of the auricle, in an aponeurosis common to other auricular muscles). However, according to authors such as Loth (1931), the temporoparietalis might be occasionally present, but is usually absent as a distinct muscle, in modern humans. According to Diogo et al. (2008, 2009b), the temporoparietalis and **auricularis anterior** derive from the auriculo-orbitalis, so when the temporoparietalis is not present as a distinct muscle, these authors

use the name auriculo-orbitalis to label the structure that is often known as 'auricularis anterior' in the literature: that is, one can only use this latter name when the temporoparietalis is present. Interestingly, the temporoparietalis was described as a distinct bundle/muscle in some gorilla specimens, seemingly corresponding, for instance, to the 'temporal superficiel' *sensu* Deniker (1885), to the 'deep portion of the auricularis anterior et superior' *sensu* Raven (1950), to a bundle/portion of the 'auricularis anterior' or 'attrahens aurem' *sensu* Bischoff (1880) and Macalister (1873), and/or to the 'attolens aurem' *sensu* Hartmann (1886). That is why Diogo et al. (2008, 2009b) stated that the temporoparietalis seems to be usually present in gorillas. However, we analyzed this feature on both sides of our VU GG1 specimen, but could not find a distinct muscle temporoparietalis: therefore, following the terminology of Diogo et al. (2008, 2009), in Figs. 2 and 3 of this specimen we prefer to label the structure that is often known as 'auricularis anterior' as auriculo-orbitalis.

- Synonymy: Auricularis anterior inferior (Raven 1950); part or totality of auricularis anterior or of attrahens aurem (Macalister 1873, Bischoff 1880, Hartmann 1886, Ruge 1887b, Gibbs 1999); auricularis anterior and possibly part or totality of pars orbito-temporalis of frontalis (Seiler 1976).

Auricularis superior (2.79 g; Figs. 2, 3)
- Usual attachments: From the superior portion of the external ear to galea aponeurotica.
- Usual innervation: Posterior auricular nerve, of CN7 (Raven 1950).
- Synonymy: Attollens aurem (Macalister 1873); part of attolens aurem (Hartmann 1886); auricularis anterior superior (Raven 1950).

Orbicularis oculi (10.31 g; Figs. 2, 3, 4)
- Usual attachments: From the nasal process of maxilla and the opposite side of the orbit, as well as from regions adjacent to these structures, to the skin near the eye, being usually divided into a pars palpebralis and a pars orbitalis, as in modern humans.
- Usual innervation: Branches of CN7 (Raven 1950).
- Synonymy: Orbiculaire des paupières (Deniker 1885); orbicularis palpebrarum (Hartmann 1886); orbicularis oculi superior et inferior (Seiler 1976).

Depressor supercilii (0.68 g; Figs. 2, 3, 4)
- Usual attachments: From the region of ligamentum palpebrale mediale to the region of eyebrow.
- Usual innervation: Branches of CN7.
- Synonymy: Orbiculaire des paupières (Deniker 1885); orbicularis palpebrarum (Hartmann 1886); orbicularis oculi superior et inferior (Seiler 1976).

Depressor supercilii (1.64 g)
- Usual attachments: From the medial region of the superciliary arch to the region of the eyebrow.

- Usual innervation: Branches of CN7.
- Synonymy: Sourcilier (Deniker 1885).

Levator labii superioris (7.52 g; Figs. 2, 3)
- Usual attachments: From the infraorbital region to the upper lip.
- Usual innervation: Branches of CN7.
- Synonymy: Releveur commun de l'aie du nez et de la lèvre supérieure (Deniker 1885); levator labii superioris proprius or maxillo-labialis (Ruge 1887a,b, Edgeworth 1935); caput infraorbitale of quadratus labii superioris (Jouffroy 1971).

Levator labii superioris alaeque nasi (1.61 g; Figs. 2, 3)
- Usual attachments: From the region of the ligamentum infraorbitale mediale to the upper lip and ala of the nose.
- Usual innervation: Branches of CN7.
- Synonymy: Releveur de l'aie du nez (Chudzinski 1885); levator alae nasi (Hartmann 1886); part or totality of naso-labialis (Edgeworth 1935); caput angulare of quadratus labii superioris (Jouffroy 1971).

Procerus
- Usual attachments: From the frontalis to the medial region of the nose.
- Usual innervation: Branches of CN7.
- Notes: According to Seiler (1971c, 1976) the '**depressor glabellae**' is usually not present as a distinct structure in gorillas, and the procerus is inconstantly present as a separate muscle in these primates.
- Synonymy: Releveur de l'aie du nez (Chudzinski 1885); levator alae nasi (Hartmann 1886); part or totality of naso-labialis (Edgeworth 1935); caput angulare of quadratus labii superioris (Jouffroy 1971).

Buccinatorius (20.82 g; Fig. 4)
- Usual attachments: From pterygomandibular raphe, infero-lateral surface of the maxilla, fossa buccinatoria and alveolar border of the mandible, to mandible, angle of the mouth, and upper and lower lips.
- Usual innervation: Branches of CN7 (Raven 1950).

Nasalis (1.36 g; Fig. 4)
- Usual attachments: From the maxilla, deep to the orbicularis oris, to the inferior portion of the lateral margin and lateral portion of inferior margin of ala of the nose.
- Usual innervation: Branches of CN7 (Raven 1950).
- Notes: Macalister (1873) stated that in the gorilla specimen he dissected there was a '**dilatator naris anterior**', but no '**dilatator naris posterior**'; however it is not clear if these structures correspond or not, to the '**dilatatores narium anterior et posterior**' *sensu* Huber (1930, 1931). According to Seiler (1970, 1971c, 1976), the '**subnasalis**' is usually not present as a distinct structure in gorillas. According to Deniker (1885) and Raven (1950), in at least some

gorillas the nasalis might be differentiated into a pars alaris ('myrtiformis' *sensu* Deniker 1885) and a pars transversa ('transversus' *sensu* Deniker 1885), as is usually the case in modern humans.

Depressor septi nasi

- Usual attachments: From the maxilla, deep to the orbicularis oris, to the inferior region of the nose.
- Usual innervation: Branches of CN7 (Raven 1950).
- Notes: Raven (1950) described a depressor septi nasi in the gorilla dissected by him. Seiler (1970, 1971c, 1976) stated that gorillas usually have a muscle depressor septi nasi, and inconstantly have a muscle '**nasalis impar**'. The depressor septi nasi described by the latter author in gorillas does seem to correspond to the depressor septi nasi he reported in other primates and that we found in taxa such as *Macaca* and *Hylobates* (see Fig. 145 of Seiler 1976 and Diogo et al. 2009b). However, the 'nasalis impar' described by Seiler in gorillas does also seem to be similar to the depressor septi nasi shown in some atlases of the modern human anatomy, being mainly a thin, infero-superiorly oriented muscle that runs from the region of the upper lip to the inferomedial region of the nose (see Fig. 145 of Seiler 1976).

Levator anguli oris facialis (2.84 g; Figs. 2, 3, 4)

- Usual attachments: From the canine fossa to the angle of mouth.
- Usual innervation: Branches of CN7 (Raven 1950).
- Notes: As proposed by Diogo et al. (2008) and Diogo and Abdala (2010), we use the name levator anguli oris facialis here (and not the name 'levator anguli oris', as is usually done in atlases of modern human anatomy) to distinguish this muscle from the **levator anguli oris mandibularis** (which is usually also labeled as 'levator anguli oris' in the literature) found in certain reptiles, which is part of the mandibular (innervated by CN5), and not of the hyoid (innervated by CN7), musculature.
- Synonymy: Caninus (Deniker 1885, Ruge 1887b, Raven 1950); part of caninus (Seiler 1976).

Orbicularis oris (10.08 g; Figs. 2, 3, 4)

- Usual attachments: From the skin, fascia and adjacent regions of the lips to skin and fascia of the lips.
- Usual innervation: Branches of CN7 (Raven 1950).
- Notes: Seiler (1970, 1971c) described a '**cuspidator oris**' in gorillas. As suggested by him, this structure, which was called '**labialis superior profundus**' by Seiler 1976, probably corresponds to the '**incisivus labii superior**' *sensu* Lightoller (1928, 1934, 1939) and, thus, to part of the orbicularis oris *sensu* the present work. Seiler (1976) also described a '**labialis inferior profundus**' in gorillas, which thus probably corresponds to the '**incisivus labii inferioris**' *sensu* Lightoller (1928, 1934, 1939) and to part of the orbicularis oris *sensu* the present work.

Depressor labii inferioris (6.56 g; Figs. 2, 3)
- Usual attachments: From the platysma myoides and the mandible to the lower lip.
- Usual innervation: Branches of CN7 (Raven 1950).
- Synonymy: Carré des lèvres (Deniker 1885); quadratus labii inferioris (Ruge 1887b, Raven 1950).

Depressor anguli oris (5.55 g; Figs. 2, 3)
- Usual attachments: From angle of the mouth to the fascia of the platysma (usually does not attach directly onto the mandible).
- Usual innervation: Branches of CN7 (Raven 1950).
- Notes: In gorillas, the depressor anguli oris is often differentiated into two or more bundles, as shown for instance in Fig. 1 of Ruge (1887b) and found in our VU GG1 specimen. In gorillas the muscle usually does not reach the inferior margin of the mandible, and does not extend inferomedially to meet its counterpart in the ventral midline, i.e., it does not form a **transversus menti**. However, Deniker (1885) stated that in the gorilla fetus dissected by him the depressor anguli oris did extend inferomedially to an aponeurosis that was probably common to the muscles of the two sides; that is, according to him the transversus menti ('transverse du menton') was present in this fetal gorilla.
- Synonymy: Triangularis (Chudzinski 1885, Deniker 1885, Ruge 1887b, Edgeworth 1935, Raven 1950); part of caninus (Seiler 1976).

Mentalis (1.35 g; Fig. 4)
- Usual attachments: From the mandible to the skin, being mainly deep to the orbicularis oris and to the depressor labii inferioris.
- Usual innervation: Branches of CN7 (Raven 1950).
- Notes: Raven (1950) stated that in the gorilla dissected by him the mentalis contacted its counterpart in the midline, but in the VU GG1 specimen dissected by us this did not seem to be the case (see Fig. 4).
- Synonymy: Levator menti (Macalister 1873); houppe du menton (Deniker 1885).

3.3 Branchial musculature

Stylopharyngeus (0.88 g; Figs. 8, 11)
- Usual attachments: This muscle has not been described in detail in gorillas. In our VU GG1 specimen it originated from the styloid process and passed between the middle and superior pharyngeal constrictors, being indirectly connected, via the connective tissue, to the greater horn of the hyoid bone.
- Usual innervation: Data not available.
- Notes: The **ceratohyoideus** and **petropharyngeus** are not present as distinct muscles in gorillas.

Trapezius (497.00 g; Figs. 5, 6, 21)
- Usual attachments: From the nuchal crest and adjacent region of the neurocranium, as well as from the vertebrae and thoracolumbar fascia, to the scapular spine, acromion and clavicle.
- Usual innervation: CN11 and often also C2, C3 and/or C4 (Deniker 1885, Eisler 1890, Sommer 1907, Raven 1950, Preuschoft 1965).
- Notes: In gorillas the trapezius usually attaches onto the lateral 1/3, or onto more than the lateral 1/3, of the clavicle, except in some cases, as for instance in our VU GG1 specimen, in which the insertion was almost onto the lateral 1/3 of the clavicle (it was onto the lateral 4.0 cm of the total 14.3 cm of this bone). As is usually the case in modern humans, in gorillas the muscle includes a pars descendens, a pars transversa, and a pars ascendens, which are however less clearly differentiated from each other than in some other primates.

Sternocleidomastoideus (199.92 g; Fig. 5)
- Usual attachments: Caput sternomastoideum from the sternum to the mastoid region and lateral portion of the nuchal crest; caput cleidomastoideum from the medial portion of the clavicle and, sometimes, also from a small portion of the sternum, to the mastoid region.
- Usual innervation: CN11 and often also C2, C3 and/or C4 (Deniker 1885, Eisler 1890, Sommer 1907, Raven 1950, Preuschoft 1965).
- Notes: The **cleido-occipitalis** is usually not present as a distinct structure in gorillas.

Constrictor pharyngis medius (0.66 g; Figs. 8, 11)
- Usual attachments: This muscle has not been described in detail in gorillas. In our VU GG1 specimen it appeared to be inserted exclusively onto the greater horn of the hyoid bone (pars ceratopharyngea; so, there was no distinct pars chondropharyngea in this specimen). The muscle (just a small part of it) seemed to be connected by the connective tissue to the neurocranium, and to contact its counterpart in the dorsal midline.
- Usual innervation: Data not available.

Constrictor pharyngis inferior (4.58 g; Figs. 6, 8)
- Usual attachments: This muscle has not been described in detail in gorillas. In our VU GG1 specimen it connected the thyroid (pars thyropharyngea, also reported by Hosokawa and Kamiya 1961-1962) and cricoid (pars cricopharyngea, also reported by Hosokawa and Kamiya 1961-1962) cartilages to the middle pharyngeal raphe.
- Usual innervation: Data not available.

Cricothyroideus (1.01 g; Figs. 6, 7, 13)
- Usual attachments: From the cricoid cartilage to the inferior and inferomesial portion, as well as to a small portion of the lateral margin of the inferior horn, of the thyroid cartilage.

- Usual innervation: Data not available.
- Notes: In our VU GG1 specimen we only found a distinct pars recta and a distinct pars obliqua; the cricothyroideus was also inserted to a small portion of the inner (inferomesial) portion of the thyroid cartilage, but we did not find a distinct pars interna such as that found in some other primates. However, Kohlbrügge (1896), Duckworth (1912), Saban (1868) and Starck and Schneider (1960) suggested that the pars interna may be differentiated in at least some gorillas. Figure 1 of Hosokawa and Kamiya (1961-1962) shows a gorilla specimen in which the muscle does not seem to meet its counterpart in the ventral midline. However, we analyzed this feature in detail in our VU GG1 specimen and the muscle did clearly meet its counterpart in the midline ventrally (see Fig. 13). The **thyroideus transversus** (found, e.g., in hylobatids) is not present as a distinct muscle in gorillas.
- Synonymy: Cricothyreoideus anticus (Kohlbrügge 1896).

Constrictor pharyngis superior (6.65 g; Fig. 11)
- Usual attachments: This muscle has not been described in detail in gorillas. In our VU GG1 specimen it connected the neurocranium and possibly the middle pharyngeal raphe to the pterygoid hamulus (pars pterygopharyngea), tongue (pars glossopharyngea, as also reported by Kleinschmidt 1938), pterygomandibular raphe (pars buccopharyngea) and mandible (pars mylopharyngea). Dean (1985) described a gorilla specimen in which the constrictor pharyngis superior originated from the anterior region of the basioccipital and also posterolaterally from the petrous apex.
- Usual innervation: Data not available.
- Notes: The **pterygopharyngeus** is not present as a distinct muscle in gorillas.

Palatopharyngeus (2.14 g; Fig. 11)
- Usual attachments: This muscle has not been described in detail in gorillas. In our VU GG1 specimen it connected the soft palate (not the pterygoid hamulus) to the pharyngeal wall and to the thyroid cartilage, and blended with the constrictor pharyngis inferior. It was not possible to discern if this specimen had a '**sphincter palatopharyngeus**' (**palatopharyngeal sphincter**, or **Passavant's ridge**), as is often the case in modern humans (see, e.g., Plate 65 of Netter 2006).
- Usual innervation: Data not available.
- Notes: On both sides of our VU GG1 specimen, we could not find fleshy fibers within the salpingopharyngeal fold; one cannot completely exclude the hypothesis that a few fibers were present, but if they were, they did not seem to form a distinct, well defined muscle **salpingopharyngeus** such as that usually found in modern humans (see, e.g., Plate 65 of Netter 2006). To our knowledge, the salpingopharyngeus has never been reported in gorillas. However, one

should keep in mind that very few authors have actually provided detailed descriptions of the pharyngeal muscles of these primates.

Musculus uvulae
- Usual attachments: This muscle has not been previously described in detail in gorillas. Edgeworth (1935) and Saban (1968) suggested that it is usually present in hominoids, but did not refer specifically to gorillas. In our VU GG1 specimen the musculus uvulae seemed to be present, originating from the soft palate, frontally (anteriorly) to the origin of the palatopharyngeus, and then running posteriorly to insert onto, or at least near to, the uvula.
- Usual innervation: Data not available.

Levator veli palatini (1.69 g; Figs. 11, 12)
- Usual attachments: This muscle has not been previously described in detail in gorillas. Dean (1985) described a gorilla in which this muscle connected the medial aspect of the eustachian process and the adjacent part of the petrous apex to the soft palate. We found similar attachments in our VU GG1 specimen. As in modern humans, in this latter specimen the muscle was somewhat medial to the tensor veli palatini, and mainly lateral to the palatopharyngeus.
- Usual innervation: Data not available.

Thyroarytenoideus (3.15 g; Fig. 14)
- Usual attachments: See notes below.
- Usual innervation: Data not available.
- Notes: There has been some controversy regarding the homologies of the thyroarytenoid bundles, as well as the presence/absence of a distinct **musculus vocalis**, within hominoids and other primates. Kohlbrügge (1896) dissected gorillas, chimpanzees, and orangutans, as well as taxa such as *Cebus*, *Semnopithecus*, *Hylobates* and *Macaca*, and stated that he could not find a distinct attachment of the thyroarytenoideus on a true vocal cord (as found in modern humans) in any of these taxa, except perhaps in *Pongo*; within all the taxa mentioned above, he found an attachment on the cricoid cartilage in *Hylobates* and *Colobus*. Giacomini (1897) examined the larynx of a gorilla and of an *Hylobates lar*, and, based on his results and on his previous work, stated that only in modern humans was there a distinct, well-developed musculus vocalis directly connected to the vocal cord, although in at least some *Hylobates*, and apparently in some *Pan* (but not in the gorilla examined by him) there were a few fibers of the thyroarytenoideus that were somewhat isolated and situated near the vocal cord (see, e.g., Fig. 2 of his Plate II). Duckworth (1912) examined specimens from all the five extant hominoid genera, as well as of *Macaca*, *Cebus*, *Semnopithecus* and *Tarsius*, and suggested that a well-developed, distinct musculus vocalis associated with the plica vocalis is only consistently present in modern humans, but great apes, and in particular chimpanzees, do show a configuration that is in some ways similar to that found in modern humans,

having for instance a distinct, but poorly developed/differentiated musculus vocalis (see, e.g., his Figs. 24 and 17). According to him, in modern humans the superior portion of the thyroarytenoideus usually also forms a distinct structure that is often associated with the region of the ventriculus (which corresponds to the '**musculus ventricularis laryngis**' *sensu* Kelemen 1948, 1969), although a similar configuration is seen in gorillas, and particularly in chimpanzees. Loth (1931) argued that in non-hominoid primates the vocal cords are mainly formed by well-developed folds of a mucous membrane, which are not really in contact with the musculus vocalis. He stated that in hominoids such as *Pan* and *Gorilla* the size of the folds is smaller, but the musculus vocalis is also not really connected to these folds; such a connection is found in most modern humans. Edgeworth (1935) argued that the musculus vocalis is found in some primates, and that, when this structure is present, the thyroarytenoideus becomes a '**thryroarytenoideus lateralis**', because its inferior/mesial part gives rise to the vocalis muscle. Starck and Schneider (1960) described such a 'pars lateralis', as well as a 'pars medialis' that usually goes to the vocal fold/cord, in *Pan*, *Pongo* and *Gorilla*, but not in *Hylobates*, the latter part thus corresponding to the musculus vocalis of modern humans. They did not find a pars aryepiglottica or a pars thyroepiglottica of the thyroarytenoideus in *Hylobates* and *Pongo*, but stated that other authors did report at least one of these structures in *Gorilla* and *Pan*. Saban (1968) clarified the nomenclature of the thyroarytenoideus, and explained that this muscle may be divided into the following structures: 1) a pars superior (often designated as '**thyroarytenoideus superior**', 'thyroarytenoideus lateralis' or '**ventricularis**'); 2) a pars inferior (often designated as '**thyroarytenoideus inferior**', '**thyroarytenoideus medialis**', or musculus vocalis); 3) a '**ceratoarytenoideus lateralis**'; 4) a 'pars intermedia' (but this name was only used by a few authors, such as Starck and Schneider (1960), who stated that some primates might have a pars superior, a pars inferior, and a pars intermedia); 5) a pars thyroepiglottica; 6) a pars aryepiglottica; 7) a pars arymembranosa; and 8) a pars thyromembranosa. According to Saban (1968), the 'ceratoarytenoideus lateralis' is usually fused with (but not differentiated from, as suggested in some anatomical atlases) the cricoarytenoideus posterior, being only a distinct muscle in a few taxa and, within primates, in *Pan* (and in this case this seems to constitute a variant/ anomaly), where this is a small muscle running from the dorsal face of the inferior thyrohyoid horn to the arytenoid cartilage. Also according to Saban (*ibid*), the relation of the pars superior and pars inferior is more superior/inferior in apes and modern humans (in modern humans the more inferior, medial part, i.e., the pars inferior, is well developed and is often labeled as musculus vocalis), and the more lateral/medial in primates such as *Macaca* and *Papio*; in *Pongo* the pars inferior is well-developed and lies anteriorly to the vocal cord, but is not associated with it. Aiello and Dean (1990) stated that in non-human hominoids the pars aryepiglottica is often reduced in size or absent. Our review of the literature indicates that: 1) the pars superior and the pars inferior of the thyroarytenoideus are present in gorillas according to Saban (1968) and Starck and Schneider (1960), the pars inferior corresponding to the musculus vocalis of humans; 2) the 'ceratoarytenoideus lateralis' and the 'pars intermedia' of the thyroarytenoideus are not present in gorillas according to Starck and Schneider (1960); 3) the pars thyroepiglottica, the pars thyromembranosa, and the pars arymembranosa of the thyroarytenoideus are present in gorillas according to Saban (1968); 4) the pars arytepiglottica of the thyroarytenoideus is present in gorillas according to Duvernoy (1855-1856) and Saban (1968). In our VU GG1 specimen, the pars aryepiglottica, pars thyroepiglottica, pars thyromembranosa, pars thyroepiglottica, 'pars intermedia', and 'ceratoarytenoideus lateralis' did not seem to be present as distinct, well-defined, structures. However, we could not discern if there was, or not, a distinct pars inferior of the thyroarytenoideus corresponding to the musculus vocalis of modern human anatomy.

Musculus vocalis (see thyroarytenoideus above)
- Usual attachments: See thyroarytenoideus above.
- Usual innervation: Data not available.
- Notes: See thyroarytenoideus above.

Cricoarytenoideus lateralis (0.47 g; Fig. 14)
- Usual attachments: From the anterior portion of cricoid cartilage to arytenoid cartilage.
- Usual innervation: Data not available.

Arytenoideus transversus (0.26 g; Fig. 14)
- Usual attachments: From the arytenoid cartilage to the contralateral arytenoid cartilage being a continuous muscle without a clear median raphe.
- Usual innervation: Data not available.
- Synonymy: Interarytenoideus (Kohlbrügge 1896).

Arytenoideus obliquus (0.08 g; Fig. 14)
- Usual attachments: See notes below.
- Usual innervation: Data not available.
- Notes: The arytenoideus obliquus was reported in gorillas by Kleinschmidt (1938), Starck and Schneider (1960) and Sonntag (1924), who stated that this muscle is well developed in these primates, being, at least in some cases, broader than the arytenoideus transversus. However, Kohlbrügge (1896) did not find the arytenoideus obliquus in the two gorillas dissected by him, and this muscle has not been described by other authors (however, one should keep in mind that the laryngeal muscles of gorillas were rarely studied in detail). In our VU GG1 specimen, we found a very thin arytenoideus obliquus, at least in one side of the body (see Fig. 14), running from the arytenoideus transversus to the region of the epiglottis and/or the tissues associated with this latter structure.

Cricoarytenoideus posterior (0.59 g; Fig. 14)
- Usual attachments: This muscle has not been described in detail in gorillas. In our VU GG1 specimen it connected the dorsal portion of the cricoid cartilage to the arytenoid cartilage; it did not meet its counterpart in the dorsal midline, and did not attach onto the inferior horn of the thyroid cartilage, i.e., there was no **ceratocricoideus** *sensu* Harrison (1995).
- Usual innervation: Data not available.

3.4 Hypobranchial musculature

Geniohyoideus (14.20 g; Figs. 11, 12)
- Usual attachments: From the mandible to the hyoid bone, lying close to its counterpart at the midline; its insertion onto the hyoid bone bifurcates the origin of the hyoglossus (see hyoglossus below).
- Usual innervation: CN12 (Raven 1950).

Genioglossus (16.58 g; Figs. 11, 12)
- Usual attachments: This muscle has not been described in detail in gorillas. Deniker (1885) stated that in the gorilla fetus dissected by him the genioglossus is separated from its counterpart close to the mandible but then connects to it, posteriorly. Hosokawa and Kamiya (1961-1962) illustrated this muscle in a gorilla specimen. In our VU GG1 specimen, the genioglossus was well separated from the geniohyoideus and also (at least its ventral portion) from its counterpart, running from the mandible to the tongue, and seemed to be loosely connected, or not connected at all, to the hyoid bone.
- Usual innervation: Data not available.
- Notes: The muscles **genio-epiglotticus**, **glosso-epiglotticus**, **hyo-epiglotticus**, and **genio-hyo-epiglotticus**, described by Edgeworth (1935) and Saban (1968) in some primate and non-primate mammals, did not seem to be present as distinct structures in our VU GG1 specimen (according to these authors, these muscles are usually not present in catarrhine taxa).

Intrinsic muscles of tongue
- Usual attachments: At least to our knowledge, there are no detailed published descriptions of these muscles in gorillas, and we were unable to analyze them in detail in our dissections. However, the **longitudinalis superior**, **longitudinalis inferior**, **transversus linguae** and **verticalis linguae** are consistently found in modern humans and at least some other primate and non-primate mammals, so these four muscles are very likely also present in gorillas. Detailed studies of the tongue and its muscles in gorillas are clearly needed.
- Usual innervation: Data not available.

Hyoglossus (3.51 g; Figs. 7, 8, 11, 12)
- Usual attachments: This muscle has not been described in detail in gorillas, although it was illustrated by Raven (1950) and Hosokawa and Kamiya (1961-1962). In our VU GG1 specimen, the hyoglossus was differentiated into a **ceratoglossus** and a **chondroglossus** (see Fig. 12), as is usually the case in modern humans (see, e.g., Terminologia Anatomica 1998). The ceratoglossus connected the greater horn of the hyoid bone to the tongue, while the chondroglossus connected the body and the small inferior horn of the hyoid bone to the tongue. These two bundles of the hyoglossus were separated by the posterolateral fibers of the geniohyoideus, which are lateral to the chondroglossus (see Fig. 12).
- Usual innervation: Data not available.

Styloglossus (3.11 g; Figs. 6, 7, 8, 11)
- Usual attachments: This muscle has not been described in detail in gorillas, although it was reported and/or illustrated by Bischoff (1880) and Dean (1984). Interestingly, in the gorilla specimen reported by Dean (1984), the styloglossus did not originate from the styloid process, but instead from the temporal bone immediately adjacent to it. In our VU GG1 specimen, the styloglossus mainly connected the styloid process (see Fig. 7) to the tongue, its fibers running longitudinally and passing laterally to the hyoglossus; only a few fibers of the styloglossus were blended with fibers of the hyoglossus.
- Usual innervation: Data not available.

Palatoglossus (1.43 g; Figs. 11, 12)
- Usual attachments: This muscle has not been described in detail in gorillas. In our VU GG1 specimen there were at least some fleshy fibers within the palatoglossal fold (Figs. 11, 12), which correspond, very likely, to the fibers of the muscle palatoglossus of modern humans and other mammals (see, e.g., Diogo et al. 2008). These fibers connected the soft palate to the supero-posterior portion of the tongue, but did not appear to extend to the lateral portion of the tongue, as is often the case in modern humans (e.g., Gibbs 1999; see also Plate 59 of Netter 2006).
- Usual innervation: Data not available.

Sternohyoideus (15.18 g; Figs. 5, 6)
- Usual attachments: From the sternum and adjacent regions to the hyoid bone.
- Usual innervation: Cervical plexus (Sommer 1907); ramus descendens of ansa hypoglossi (C1, C2, C3; Raven 1950).
- Notes: In the gorilla fetus dissected by Deniker (1885) the sternohyoideus did not have tendinous intersections such as those that are found in at least some modern humans, chimpanzees, hylobatids and other primate and non-primate mammals. In our VU GG1 specimen we also did not find such tendinous intersections. Deniker (1885) stated that in the gorilla fetus analyzed by him

the sternohyoideus did not contact to its counterpart in the midline, but in the adult gorillas described by Duvernoy (1855-1856) and Raven (1950), as well as in our VU GG1 specimen, the sternohyoideus clearly contacted to its counterpart, however being well separated from it anteriorly, near its insertion onto the hyoid bone.

Omohyoideus (4.70 g; Figs. 5, 6, 8)

- Usual attachments: From the scapula to the hyoid bone.
- Usual innervation: Cervical plexus (Sommer 1907); branch of ansa hypoglossi (Raven 1950).
- Notes: Among gorillas, a distinct, well-defined intermediate tendon of the omohyoideus was reported in one specimen by Macalister (1873) and one specimen by Raven (1950), but was missing in two specimens dissected by Deniker (1885), one specimen reported by Duvernoy (1855-1856), one specimen reported by Bischoff (1880), and one specimen examined by Testud (1884). In the specimen described by Macalister (1873) there was a superior belly and an inferior belly, as is usually the case in modern humans, but in the specimen dissected by Raven (1950) the omohyoideus had three bellies on the left side and two bellies on the right side. As explained in the recent work of Rajalakshmi et al. (2008), the presence of three bellies of the omohyoideus is not an uncommon anatomical variant in modern humans and according to their review found in about 3% of modern human subjects. Interestingly, in at least some chimpanzees the omohyoideus also has three bellies (e.g., Sonntag 1924; our dissections). Regarding our VU GG1 specimen, the right side of the omohyoideus had a single belly, with no intermediate tendon or tendinous intersection (see Fig. 6), running from the hyoid bone to the anteromesial portion of the scapula, just laterally to the insertion of the levator claviculae; on the left side the attachments were similar, but there was a small tendinous intersection (but not a distinct, well-defined intermediate tendon as is usually found in modern humans) in the middle of the muscle (see Fig. 8).

Sternothyroideus (3.19 g; Fig. 6)

- Usual attachments: From the sternum and adjacent regions to the oblique line of the thyroid cartilage.
- Usual innervation: Cervical plexus (Sommer 1907); branches of ansa hypoglossi (C1, C2, C3; Raven 1950).
- Notes: Figure 20 of Raven (1950) seems to suggest that in the gorilla dissected by him the sternothyroideus had a tendinous intersection such as those present in at least some modern humans, chimpanzees, hylobatids, and some other primate and non-primate taxa. Fürbringer (1872) reported such an intersection in another gorilla specimen, but Deniker (1885) stated that there was no tendinous intersection in the gorilla fetus dissected by him. In our VU GG1 specimen the sternothyroideus runs from the sternum to the thyroid

cartilage: on the right side of the body a few fibers were fused with the fibers of the thyrohyoideus, but on the left side no fibers were fused with this latter muscle. The insertion was mainly (right side) or exclusively (left side) onto the thyroid cartilage, from the posterolateral surface to a more anterior portion of this cartilage, passing anteriorly to the posterior portion of the thyrohyoideus. The sternothyroideus was well separated from its counterpart in the midline and from the sternohyoideus, and did not have a tendinous intersection.

Thyrohyoideus (1.87 g; Fig. 6)
- Usual attachments: From the thyroid cartilage to the hyoid bone.
- Usual innervation: Cervical plexus (Sommer 1907); branches of ansa hypoglossi (C1, C2, C3; Raven 1950).

3.5 Extra-ocular musculature

Muscles of eye
- Usual attachments: To our knowledge, there are no detailed published descriptions of these muscles in gorillas, and we could not analyze them in detail in our dissections. However, the **orbitalis, rectus inferior, rectus superior, rectus medialis, rectus lateralis, levator palpebrae superioris, obliquus superior** and **obliquus inferior** are consistently found in modern humans and at least some other primate and non-primate mammals, so these muscles are also likely to be present in gorillas. Detailed studies of the eye and its muscles in gorillas are clearly needed.
- Usual innervation: Data not available.

CHAPTER 4

Pectoral and Upper Limb Musculature

Serratus anterior (526.0 g; Figs. 15, 16, 18, 21, 22, 53, 54, 55)
- Usual attachments: From the ribs to the medial border of the scapula, being well separated from the levator scapulae.
- Usual innervation: Long thoracic nerve (Deniker 1885, Eisler 1890, Sommer 1907, Preuschoft 1965), from C5 and C6 according to Hepburn (1892).
- Notes: In gorillas, the serratus anterior might originate from ribs 1–13 (e.g., specimens of Hepburn 1892 and Raven 1950 and the juvenile specimen of Deniker 1885), ribs 1–12 (e.g., specimen of Stewart 1936 and fetal specimen of Deniker 1885), ribs 1–11 (e.g., specimens of Duvernoy 1855-1856 and Bischoff 1880), or ribs 1–10 (e.g., specimen of Macalister 1873).
- Synonymy: Grand dentelé (Deniker 1885); serratus magnus (Macalister 1873, Hepburn 1892).

Rhomboideus (179.0 g; Figs. 21, 22)
- Usual attachments: From the cervical and thoracic vertebrae to the medial border of the scapula.
- Usual innervation: Dorsal scapular nerve, from C3, C4 and/or C5 (Eisler 1890, Hepburn 1892, Sommer 1907, Preuschoft 1965), and sometimes also by the intercostal nerves 3 and 4 according to Eisler (1890).
- Notes: The **rhomboideus major**, **rhomboideus minor** and **rhomboideus occipitalis** are not present as distinct structures in gorillas. In these primates the rhomboideus usually extends anteriorly to C5 (e.g., specimen of Macalister 1873), C4, C3 (e.g., specimen of Raven 1950 and one specimen of Preuschoft 1965), C2 (two specimens of Preuschoft 1965), or even C1, and extends posteriorly to T2 (e.g., juvenile specimen of Deniker 188), T3, T4 (e.g., specimens of Macalister 1873 and Stewart 1936, fetal specimen of Deniker 1885, and one specimen of Preuschoft 1965), T5 (two specimens of Preuschoft 1965) or even T6 (specimen of Raven 1950). In our VU GG1 specimen it seemed to originate from C1–T5.

Levator scapulae (68.6 g; Figs. 6, 22)
- Usual attachments: From cervical vertebrae to the medial border of the scapula.
- Usual innervation: Dorsal scapular nerve, from C3, C4 and/or C5 (Eisler 1890, Raven 1950, Preuschoft 1965); dorsal scapular nerve and cervical plexus (Sommer 1907).
- Notes: In gorillas, the levator scapulae usually originates from cervical ribs C1–C4 (e.g., specimens of Hepburn 1892, Pira 1913, Raven 1950 and Preuschoft 1965 and fetal specimen of Deniker 1885) or C1–C5 (e.g., specimens of Duvernoy 1855-1856, Sommer 1907 and Stewart 1936). Macalister (1873) stated that in the gorilla dissected by him the muscle arose from C3–C6 or but this description is somewhat doubtful, because in all the other gorilla specimens, as well as in the vast majority of other hominoids, the levator scapulae usually originates partially from C1 and/or C2. In our VU GG1 specimen the origin of the levator scapulae was from C1–C4.
- Synonymy: Angulaire de l'omoplate (Deniker 1885); levator anguli scapulae (Macalister 1873, Hepburn 1892); levator scapulae dorsalis (Jouffroy 1971).

Levator claviculae (22.7 g; Fig. 5)
- Usual attachments: From the atlas to the clavicle.
- Usual innervation: C2 and C3 (Eisler 1890), C3 and C4 (Sommer 1907), C4 and C5 (Preuschoft 1965), or C2, C3 and C4 (Raven 1950).
- Notes: In gorillas, the levator claviculae goes exclusively to the clavicle, except in very few cases, e.g. the specimens reported by Preuschoft (1965), in which the muscle is attached to the clavicle and scapula. Andrews and Groves (1976) stated that in gorillas the insertion of the muscle onto the clavicle is superficial/lateral to that of the trapezius, but Deniker (1885), Sommer (1907), Schück (1913a,b), Raven (1950), Preuschoft (1965) and Jouffroy (1971) stated that it is actually deep/medial to the this latter muscle, a statement that was corroborated by our dissections (see Fig. 5). Stewart (1936) stated that the attachment of the levator claviculae onto the clavicle extended medially to a point just lateral to the midpoint of this bone, and in our VU GG1 specimen it actually reached the midpoint of this bone (see Fig. 5). The **atlantoscapularis posticus** (see, e.g., Diogo et al. 2009a) is usually not present as a distinct muscle in gorillas.
- Synonymy: Omocervicalis, cleidocervicalis, acromio-cervicalis, acromiotrachealis, clavotrachealis, cleido-omo-transversalis, tracheloacromialis or levator anticus scapulae (Duvernoy 1855-1856, Bischoff 1880, Deniker 1885, Macalister 1873, Barnard 1875, Sommer 1907, Preuschoft 1965); omo-cleido-transversarius or cleido-atlanticus (Pira 1913); atlantoclavicularis (Raven 1950, Aiello and Dean 1990); atlantoscapularis anterior (Ashton and Oxnard 1963); levator scapulae ventralis (Jouffroy 1971).

Subclavius (4.3 g; Fig. 17)
- Usual attachments: From the first rib to the clavicle.
- Usual innervation: Nerve to subclavius (Eisler 1890, Hepburn 1892, Sommer 1907, Preuschoft 1965), from C5 and C6 according to Hepburn (1892), although Eisler (1890) and Preuschoft (1965) also refer to an innervation by pectoral nerves.
- Notes: Duvernoy (1855-1856) stated that he did not find the subclavius in the gorilla specimen dissected by him. Preuschoft (1965) stated that in the gorillas examined by him the muscle originated from the first rib and also from the sternoclavicular joint, and that in one of these specimens the muscle inserted onto the clavicle but also onto the coracoid process of the scapula. The **costocoracoideus** (see, e.g., Diogo et al. 2009a) is not present as a distinct muscle in gorillas, but these primates do usually have a ligamentum costocoracoideum (see Fig. 18).
- Synonymy: Sous-clavier (Deniker 1885).

Pectoralis major (651 g; Figs. 15, 16, 17, 18, 53, 54)
- Usual attachments: Pars clavicularis from the medial portion of the clavicle and the anterior portion of the sternum to the proximal portion of the humerus; pars sternocostalis from the sternum and ribs to the proximal portion of the humerus, proximally to the insertion of the pars clavicularis; pars abdominalis from the ribs and aponeurosis of external oblique to the coracoid process of the scapula, being deep to the pars sternocostalis and pars clavicularis.
- Usual innervation: Medial pectoral nerve, from C7, and lateral pectoral nerve, from C6 and C7 (Hepburn 1892); anterior/ventral thoracic (pectoral) nerves (Sommer 1907, Preuschoft (1965).
- Notes: Due to its usual attachment onto the coracoid process of the scapula, various authors have called the pars abdominalis of the pectoralis major of gorillas under the name '**pectoralis quartus**' or '**pectoralis abdominis**', suggesting that this structure is not homologous to the pars abdominalis of the pectoralis major of modern humans (which usually goes to the humerus and not to the coracoid process of the scapula), and that it is possibly derived from the 'entopectoralis' *sensu* Lander 1918 (for a recent review on the homologies of the 'entopectoralis' and of the 'ectopectoralis', see Diogo et al. 2009a). However, a detailed comparison with modern humans and other primate and non-primate mammals clearly seems to indicate that this structure does correspond to the pars abdominalis of modern humans (and of other primates), because in gorillas this portion of the pectoralis major lies deep, and posteriorly, to the pars sternocostalis, mainly from the ribs and from the external oblique, as is the case in modern humans. Also, it is clearly more parsimonious, in a cladistic context, to consider that the gorilla and modern human structures are homologous, than to assume that the pars abdominalis was secondarily lost in gorillas (because this structure is also present in *Pan*, *Pongo,* and *Hylobates*), and that, at the same time, gorillas acquired a

new, and very similar, structure (i.e., a 'pectoralis quartus' or 'pectoralis abdominis'). The homology between the structure of gorillas and the structure found in modern humans and other primates was also defended by Hepburn (1892) and Stewart (1936). Interestingly, Miller (1932) stated that in gorillas the pars abdominalis of the pectoralis major might have an additional origin from the latissimus dorsi. According to Deniker (1885), Duckworth (1898), Miller (1932), Stewart (1936), Raven (1950) and Preuschoft (1965), in gorillas the pectoralis major is usually deeply blended with the biceps brachii; this was effectively the case in our VU GG2 specimen (see Fig. 18), but not in our VU GG1 specimen (see Fig. 17). According to Loth (1931), in 100% of gorillas there is contact between the pectoralis major and its counterpart at the ventral midline, and this was also corroborated by Stewart (1936), although in the fetal gorilla described by Deniker (1885) there was a distance of 1mm between the muscles of the two sides of the body.

- Synonymy: Pectoralis major plus lower portion of pectoralis minor (Hartmann 1886); pectoralis major plus pectoralis abdominis, pectoralis abdominalis, pectoralis quartus and/or chondroepitrochlearis quartus (Sonntag 1924, Miller 1932, Raven 1950, Aiello and Dean 1990, Gibbs 1999).

Pectoralis minor (76.0 g; Figs. 16, 17, 18)
- Usual attachments: From the ribs to the coracoid process of the scapula.
- Usual innervation: Medial pectoral nerve, from C7 (Hepburn 1892); anterior/ventral thoracic (pectoral) nerves (Deniker 1885, Eisler 1890, Sommer 1907, Stewart 1936, Raven 1950, Preuschoft 1965).
- Notes: The **pectoralis tertius** ('**xiphihumeralis**'), **sternalis**, **supracostalis** and **panniculus carnosus** (see, e.g., Diogo et al. 2009a) are usually not present as distinct muscles in gorillas, although Jouffroy (1971) stated that a 'vestigial' panniculus carnosus might be occasionally found in these primates.
- Synonymy: Upper portion of pectoralis minor (Hartmann 1886).

Infraspinatus (244.9 g; Fig. 22)
- Usual attachments: From the infraspinous fossa of the scapula and infraspinatus fascia to the greater tuberosity of the humerus and capsule of the glenoid joint.
- Usual innervation: Suprascapular nerve (Eisler 1890, Hepburn 1892, Sommer 1907, Raven 1950, Preuschoft 1965), from C4 and C5 according to Hepburn (1892).

Infraspinatus (180.4 g; Fig. 22)
- Usual attachments: From the supraspinous fossa of the scapula and supraspinatus fascia to the greater tuberosity of the humerus and the capsule of the glenoid joint.

- Usual innervation: Suprascapular nerve (Eisler 1890, Hepburn 1892, Sommer 1907, Raven 1950, Preuschoft 1965), from C4 and C5 according to Hepburn (1892).
- Notes: The muscle **scapuloclavicularis**, occasionally present in modern humans, was not described in gorillas nor found in the gorilla specimens dissected by us.

Deltoideus (705.2 g; Figs. 15, 16, 18, 19, 21, 22, 53)
- Usual attachments: From the lateral portion of the clavicle (pars clavicularis), acromion (pars acromialis) and scapular spine and infraspinatus fascia (pars spinalis) to the middle third of the humerus.
- Usual innervation: Axillary ('circumflex') nerve (Deniker 1885, Eisler 1890, Hepburn 1892, Sommer 1907, Raven 1950, Preuschoft 1965).

Teres minor (68.3 g; Fig. 22)
- Usual attachments: From the infraspinatus fascia and lateral border of the scapula to the greater tuberosity of the humerus, often also extending distally to this tuberosity.
- Usual innervation: Axillary ('circumflex') nerve (Eisler 1890, Hepburn 1892, Sommer 1907, Raven 1950, Preuschoft 1965).
- Notes: In gorillas the teres minor is usually well separated from the infraspinatus. An insertion of the teres minor onto the shaft of the humerus, distally to the greater tuberosity of this bone, was reported by, e.g., Raven (1950), Gibbs (1999) and Gibbs et al. (2002) and was also found in our VU GG1 specimen, this also seems to be the case in the gorillas illustrated by Preuschoft (1965).
- Synonymy: Petit rond (Deniker 1885).

Subscapularis (449.8 g; Fig. 20)
- Usual attachments: From the subscapular fossa of scapula to the lesser tuberosity of the humerus, occasionally also extending distal to this tuberosity.
- Usual innervation: Subscapular nerves (Eisler 1890, Hepburn 1892, Sommer 1907, Raven 1950, Preuschoft 1965). Eisler (1890) and Hepburn (1892) stated that in the gorillas dissected by them at least part of these nerves arose from the axillary nerve, and then gave rise to two branches, one going to the subscapularis muscle, and the other going to both this muscle and the teres major.
- Notes: An insertion of the subscapularis onto the shaft of the humerus, distally to the lesser tuberosity of this bone, was present in our VU GG1 specimen, but was not reported by Raven (1950), Preuschoft (1965), Gibbs (1999) and Gibbs et al. (2002).

Teres major (346.7 g; Figs. 20, 21, 22)
- Usual attachments: From the lateral border and inferior angle of the scapula to the proximal portion of the humerus, by means of a distal tendon that passes dorsally to the distal tendon of the latissimus dorsi (see Fig. 22).

- Usual innervation: Subscapular nerves (Eisler 1890, Hepburn 1892, Sommer 1907, Raven 1950, Preuschoft 1965). As explained above, Eisler (1890) and Hepburn (1892) stated that in the gorillas dissected by them at least part of these nerves arose from the axillary nerve, and then gave rise to two branches, one going to the subscapularis muscle, and the other going to both this muscle and the teres major. In one specimen and one side of the other specimen dissected by Preuschoft (1965) the teres major was innervated by the axillary nerve, while in the other side of the latter specimen it was innervated by the subscapular nerves; in a third specimen dissected by this author the muscle was innervated by the axillary nerve and by the thoracodorsal (middle subscapular) nerve.
- Synonymy: Grand rond (Deniker 1885).

Latissimus dorsi (783.0 g; Figs. 20, 21, 22)
- Usual attachments: From the vertebrae, ribs, thoracolumbar fascia and, often, directly and/or indirectly from the pelvis, to the proximal portion of the humerus.
- Usual innervation: Thoracodorsal (middle subscapular) nerve (Deniker 1885, Eisler 1890, Hepburn 1892, Sommer 1907, Raven 1950, Preuschoft 1965).
- Notes: As noted by Raven (1950), Preuschoft (1965) and Gibbs (1999), and corroborated in our VU GG1 specimen, in gorillas the posteromedial portion of the trapezius is deeply blended with the anteromedial portion of the latissimus dorsi, as is also often the case in chimpanzees, orangutans and some other primates, but not in modern humans.
- Synonymy: Grand dorsal (Deniker 1885).

Dorsoepitrochlearis (39.6 g; Figs. 20, 22)
- Usual attachments: From the distal portion of the latissimus dorsi to the medial epicondyle of the humerus.
- Usual innervation: Radial nerve (Deniker 1885, Eisler 1890, Hepburn 1892, Sommer 1907, Pira 1913, Raven 1950, Preuschoft 1965), by a ramus from the same branch that innervates the long head of the triceps brachii according to Raven (1950).
- Notes: As noted by Raven (1950), in the specimen dissected by him the dorsoepitrochlearis originated from the latissimus dorsi and from a flat aponeurotic tendon connected to the axillary border of the scapula, in common with the long head of the triceps brachii.
- Synonymy: Latissimo-condylus or latissimo-epitrochlearis (Barnard 1875); latissimo-condyloideus (Chapman 1878, Hepburn 1892, Sommer 1907); latissimo-tricipitalis (Preuschoft 1965).

Triceps brachii (749.7 g; Figs. 19, 20, 22, 30)
- Usual attachments: From the lateral portion of the lateral border of the scapula (caput longum) and shaft of the humerus (caput laterale often extending, proximally, near the surgical neck of the humerus, e.g., it lies just 0.5 cm distal to this structure in our VU GG1 specimen; caput mediale often lying far from

this surgical neck, e.g., it lies 15 cm distal to the proximal extremity of the humerus in our VU GG1 specimen, in which the total length of the humerus is about 36 cm), to the olecranon process of the ulna.

- Usual innervation: Radial nerve (Eisler 1890, Hepburn 1892, Sommer 1907, Raven 1950, Preuschoft 1965), but also by branches of the ulnar and/or axillary nerves according to Preuschoft (1965).

- Notes: Deniker (1885) stated that in the fetal gorilla dissected by him the long head of the triceps brachii had a double origin from the scapula, as described by Duvernoy (1855-1856). Payne (2001) described two gorillas with a particularly well-developed triceps brachii, which had an unusual insertion: it crossed superficial to the origin of the forearm extensors to insert approximately 4 inches distally to the olecranon process of the ulna. Loth (1931) and Preuschoft (1965) state that in gorillas the long head of the triceps originates from the distal 1/2 of the lateral border of the scapula, and Gibbs (1999) refers to the distal 1/3 to 1/2 of this border; however, in our VU GG1 specimen, the origin was only from the distal 5 cm (of the total 18 cm) of the lateral border of the scapula, i.e., it was from less than 1/3 of this border (see Fig. 22).

- Synonymy: Multiceps extensor cubiti (Barnard 1875); triceps extensor cubiti (Hepburn 1892).

Brachialis (314.9 g; Figs. 15, 19, 20, 23)
- Usual attachments: From the humeral shaft distal to the surgical neck of the humerus (e.g., it is 15 cm distal to the proximal extremity of the humerus in our VU GG1 specimen, in which the total length of the humerus is ca. 36 cm), to the ulnar tuberosity, and sometimes also to the capsule of elbow joint and/or to the coronoid process of the ulna.

- Usual innervation: Musculocutaneous nerve (Eisler 1890, Hepburn 1892, Sommer 1907, Raven 1950, Preuschoft 1965, Koizumi and Sakai 1995), but occasionally also by branches of the radial nerve according to Raven (1950) and Preuschoft (1965), and by branches of both the radial and median nerves according to Hepburn (1892).

- Notes: An insertion onto the capsule of the elbow joint and the coronoid process of the ulna was described by Raven (1950).

- Synonymy: Brachialis anticus or brachialis anterior (Owen 1868, Chapman 1878, Deniker 1885, Hepburn 1892).

Biceps brachii (402.6 g; Figs. 15, 17, 18, 19, 20, 23)
- Usual attachments: From the coracoid process (caput breve) and the supraglenoid tubercle (caput longum) of the scapula, to the bicipital tubercle of the radius (common tendon) and to the fascia lying over the forearm flexors (bicipital aponeurosis, or 'lacertus fibrosus').

- Usual innervation: Musculocutaneous nerve (Eisler 1890, Hepburn 1892, Höfer 1892, Sommer 1907, Raven 1950, Preuschoft 1965, Koizumi and Sakai 1995).
- Notes: Among gorillas, the usual condition seems to be that in which the bicipital aponeurosis is present, as described by Hartmann (1886), Sommer (1907), Pira (1913), Raven (1950) and Preuschoft (1965) and also found in our VU GG1 specimen (see Fig. 23), although this structure is apparently lacking in a few specimens, such as those described by Owen (1868) and Duvernoy (1855-1856). Sommer (1907) stated that he found some muscular fibers associated with the bicipital aponeurosis of the gorilla dissected by him (in this case, one refers to a 'lacertus carnosus', and not to a 'lacertus fibrosus'), but such muscular fibers were not found in the gorillas dissected by other authors or by us.
- Synonymy: Biceps flexor cubiti (Owen 1868, Hepburn 1892).

Coracobrachialis (67.7 g; Figs. 17, 20, 23)
- Usual attachments: From the coracoid process to the proximal portion of the humerus.
- Usual innervation: Musculocutaneous nerve (Eisler 1890, Hepburn 1892, Höfer 1892, Sommer 1907, Raven 1950, Preuschoft 1965, Koizumi and Sakai 1995), although Koizumi and Sakai (1995) stated that this was not the case in 1/3 of the gorillas dissected by them.
- Notes: In the gorilla dissected by Hepburn (1892) the coracobrachialis presented variations on the two sides of the body. On the right side it arose from the tip of the coracoid process of the scapula and passed downwards and backwards to be inserted into the middle third of the inner surface of the shaft of the humerus. From the lower border of the muscle a few fibrous strands passed downwards superficial to the musculocutaneous nerve and became attached to the internal intermuscular septum; according to Hepburn (1892) these strands probably represent the 'coracobrachialis longus'. However, as stressed by Howell and Straus 1932, the strands described by Hepburn actually seem to represent an extension of the **coracobrachialis medius/coracobrachialis proprius**, i.e., they do not correspond to the **coracobrachialis superficialis/coracobrachialis longus** *sensu* Diogo et al. 2009b. In addition to the parts described above, on the left side of the gorilla a few muscular fibers connected the under surface of the coracobrachialis to the inner side of the neck of the humerus above the level of the teres major tendon; according to Hepburn (1892), these latter muscular fibers correspond to the **coracobrachialis profundus/coracobrachialis brevis** of other primates (see Diogo et al. 2009a). In the vast majority of the gorillas dissected by other authors and by us, the coracobrachialis brevis is not present as a distinct structure. Contrary to other hominoids and to most other primates, in gorillas (and also in hylobatids) the usual condition seems to be that in which the musculocutaneous nerve does not pass through the coracobrachialis, as noted by Macalister (1873), Hepburn (1892), Sonntag (1924), Loth (1931),

Raven (1950) and Gibbs (1999) and illustrated in Fig. 4 of Preuschoft (1965), although the nerve may pierce the muscle in a few gorilla specimens (e.g., Chapman 1878; this was also the case in our specimen VU GG1).

- Synonymy: Biceps flexor cubiti (Owen 1868, Hepburn 1892).

Pronator quadratus (13.3 g; Figs. 29, 46)

- Usual attachments: From the distal portion of the ulna to the distal portion of the radius.
- Usual innervation: Median (anterior interosseous) nerve (Eisler 1890, Hepburn 1892, Sommer 1907, Raven 1950, Preuschoft 1965).
- Synonymy: Carré pronateur (Deniker 1885).

Flexor digitorum profundus (336.8 g; Figs. 24, 27, 28, 38, 40, 45, 46, 47)

- Usual attachments: From the radius, ulna, and interosseous membrane to the distal phalanges of digits 2, 3, 4 and 5, and, often, by means of a 'vestigial', thin tendon, of digit 1, although this tendon is missing in some specimens.
- Usual innervation: Median and ulnar nerves (Eisler 1890, Hepburn 1892, Sommer 1907, Pira 1913, Raven 1950, Preuschoft 1965); Pira (1913) and Hepburn (1892) stated that the ulnar nerve innervated the portions to digits 4 and 5, while Sommer (1907) stated that this nerve innervated the portions to digits 3, 4 and 5.
- Notes: According to Keith (1894b), the tendon to digit 1 is completely absent in about 2/3 (8 out of 12) gorillas. According to the review done by Straus (1942), this structure was lacking in 31% (5 out of 16), rudimentary and inutile in 38% (6 out of 16), and 'completely present' in 25% (4 out of 16) of the cases, in gorillas. Among the four hands of the two gorillas in which we could analyze this feature in detail, VU GG1 and CMS GG1, this tendon was present, but was a 'vestigial', thin structure. In fact, in these gorillas, as well as in a great part of the gorillas dissected by other authors in which the tendon to digit 1 is present, this tendon mainly originated from the fascia lying over the thenar muscles of the hand. That is, it does not really reach, proximally, the tendons of the flexor digitorum profundus to digits 2–5 (see Figs. 28, 45), and is not associated with a separate, well-developed fleshy belly, as is usually the case in modern humans and hylobatids; i.e., contrary to these two latter groups, there is usually no distinct **flexor pollicis longus** in gorillas. Interestingly, Raven (1950) stated that in the gorilla dissected by him there was a separate flexor pollicis longus, with a belly that was distinct from, and only connected by a small tendon (beneath the flexor retinaculum) to, the main belly of the flexor digitorum profundus, originating from the radius and interosseous membrane and inserting, via a thin tendon, to the distal phalanx of the thumb.

- Synonymy: Flexor profundus (Owen 1868); flexor digitorum communis profundus (Barnard 1875); flexor profundus digitorum (Chapman 1878, Hartmann 1886); fléchisseur profond (Deniker 1885).

Flexor digitorum superficialis (226.9 g; Figs. 23, 24, 25, 26, 27, 37a,b, 38a,b, 39, 40, 41, 42)

- Usual attachments: From the radius (caput radiale) and the ulna and the medial epicondyle of the humerus (caput humeroulnare) to the middle phalanges of digits 2–5.
- Usual innervation: Median nerve (Eisler 1890, Hepburn 1892, Höfer 1892, Sommer 1907, Raven 1950, Preuschoft 1965).
- Notes: Hepburn (1892), Loth (1931), Raven (1950), Preuschoft (1965), Jouffroy (1971) and Gibbs (1999) refer to an origin including the ulna, radius and medial epicondyle of the humerus and/or the common flexor tendon, but there was no ulnar origin in Deniker's (1885) fetal gorilla, in Macalister (1873) specimen, and in our VU GG1 specimen. In the gorilla specimens described by Owen (1868), Macalister (1873), Hepburn (1892), Raven (1950), Sarmiento (1994) the **palmaris longus** was present, but the palmaris longus was absent in, e.g., the specimens reported by Chapman (1878), Bischoff (1880), Deniker (1885), Hartmann (1886), Symington (1889), Eisler (1890), Hepburn (1892), Höfer (1892), Duckworth (1904), Sommer (1907), Pira (1913) and in 5 of the 6 upper limbs dissected by Preuschoft (1965), as well as in both sides of our VU GG1 specimen. In the literature reviews done by other authors, the palmaris longus is present in 7 out of 11 cases according to Keith (1899), in 15% of the cases according to Loth (1931), in 4 out of 11 cases according to Sarmiento (1994), and in 6 out of 19 cases according to Gibbs (1999). Therefore, it can be said that the palmaris longus is usually missing in gorillas.
- Synonymy: Flexor sublimis digitorum (Owen 1868, Hartmann 1886, Hepburn 1892); fléchisseur superficiel (Deniker 1885); flexor digitorum sublimis (Sommer 1907); flexor digitorum sublimis plus flexor indicis proprius (Raven 1950).

Flexor carpi ulnaris (85.0 g; Figs. 23, 25, 30, 37)

- Usual attachments: From the ulna (caput ulnare) and medial epicondyle of the humerus (caput humerale) to the pisiform.
- Usual innervation: Ulnar nerve (Eisler 1890, Hepburn 1892, Höfer 1892, Sommer 1907, Raven 1950, Preuschoft 1965).
- Notes: The **epitrochleoanconeus** was not found or described in the gorillas reported by Duvernoy (1855-1856), Macalister (1873), Chapman (1878), Bischoff (1880), Deniker (1885), Hartmann (1886), Eisler (1890), Hepburn (1892), Höfer (1892), Sommer (1907), Pira (1913), Raven (1950) and Preuschoft (1965) and those dissected by us.
- Synonymy: Cubital antérieur (Deniker 1885).

Flexor carpi radialis (89.8 g; Figs. 23, 25, 30, 37)

- Usual attachments: From the radius and medial epicondyle of the humerus to bases of metacarpals II and III.
- Usual innervation: Median nerve (Eisler 1890, Hepburn 1892, Höfer 1892, Sommer 1907, Raven 1950, Preuschoft 1965).
- Notes: Among gorillas, an insertion on the metacarpals II and III was described by Hepburn (1892), Raven (1950) and Preuschoft (1965; this author also referred to an insertion onto the trapezius), and according to Jouffroy (1971) and Gibbs (1999) such an insertion on these two metacarpals is usually, and possibly always, found in gorillas; we did effectively find this configuration in our dissections.
- Synonymy: Grand palmaire (Deniker 1885; this author confused the flexor carpi radialis with the palmaris longus, which is usually absent in gorillas: see above).

Pronator teres (81.7 g; Figs. 23, 25, 30)

- Usual attachments: From the humerus (caput humerale) and, often (see notes below), also from the ulna to the radius.
- Usual innervation: Median nerve (Eisler 1890, Hepburn 1892, Höfer 1892, Sommer 1907, Raven 1950, Preuschoft 1965). When both the caput humerale and caput ulnare are present, this nerve often passes between these heads; when the caput ulnare is missing, the nerve usually passes dorsally to the muscle.
- Notes: Owen (1868) stated that in gorillas the caput humerale and caput ulnare are present, but are not as well defined as in modern humans, and in the specimens of Duvernoy (1855-1856), the specimen of Chapman (1878), the specimen of Bischoff (1880), two of the three specimens of Hartmann (1886), one of the two specimens of Deniker (1885), the specimen of Eisler (1890), the specimen of Hepburn (1892), the specimen of Höfer (1892) and the specimen of Lewis (1989), as well as in our VU GG1 specimen, there was no distinct ulnar head. However, in the specimen of Macalister (1873), one of the three specimens of Hartmann (1886), one of the two specimens of Deniker (1885), the specimen of Symington (1889), the specimen of Sommer (1907), the specimen of Pira (1913), the specimen of Miller (1932), and the specimen of Raven (1950), the two heads were present. According to Parsons (1898b), Chylewski (1925), Loth (1931), Preuschoft (1965) and Jouffroy (1971) the two heads are present in 40%, 44%, 44%, 47% and 42% of gorillas, respectively. But if one takes into account the extensive review of the literature done by Preuschoft (1965) plus the presence of the two heads in the specimen of Owen (1868) and in the three specimens of Preuschoft (1965) and its absence in the specimen of Lewis (1989) and in our VU GG1 specimen, one can conclude that the two heads are present in 12 out of 23 gorillas, i.e., in 52% of the cases.
- Synonymy: Pronator radii teres (Chapman 1878, Hepburn 1892); rond pronateur (Deniker 1885).

Palmaris brevis (0.3 g; Figs. 39, 41)

- Usual attachments: From the pisiform and/or flexor retinaculum to the skin of the median border of the palm.
- Usual innervation: Superficial branch of the ulnar nerve (Raven 1950, Preuschoft 1965).
- Notes: Among gorillas, the palmaris brevis was not reported in the specimen of Bischoff (1880), the fetus of Deniker (1895), the specimen of Hepburn (1892), the specimen of Duckworth (1904), the specimen of Pira (1913), and three of the four specimens of Preuschoft (1965), but it was present in the specimen of Duvernoy (1855-1856), the juvenile specimen of Deniker (1885), the specimen of Höfer (1892), the specimen of Sommer (1907), the specimen of Raven (1950), one of the four specimens of Preuschoft (1965), the specimen illustrated by Dylevsky (1967), the two specimens of Sarmiento (1994), our CMS GG1 specimen, and in at least one side of our VU GG1 specimen (see Figs. 39, 41). According to the literature reviews done by Sarmiento (1994) and Gibbs (1999) the muscle is present in 7/9 and 1/2 gorillas, respectively, and according to our own review of the literature plus the data obtained in our own dissections, it is present in 11/19 (58%) of the cases. The **flexor digitorum brevis manus** and **palmaris superficialis** (see, e.g., Diogo et al. 2009a) are usually not present as distinct muscles in gorillas.
- Synonymy: Palmaire cutané (Deniker 1885).

Lumbricales (I: 2.9 g; II: 2.9 g, III: 2.5 g; IV: 0.9 g; Figs. 24, 37, 38, 39, 40, 41, 42, 43, 44)

- Usual attachments: From the tendons of the flexor digitorum profundus to digits 2, 3, 4 and 5 (an origin from the tendon to digit 1 was almost never found, except by, e.g., Raven 1950) to the radial side of the proximal phalanx and of extensor expansion of digits 2 (lumbricalis I), 3 (lumbricalis II), 4 (lumbricalis III) and 5 (lumbricalis IV).
- Usual innervation: Hepburn (1892) stated that in the gorilla dissected by him the lumbricalis I was innervated by the median nerve, the II by the median nerve and the deep branch of the ulnar nerve, and the III and IV by the deep branch of the ulnar nerve. Sommer (1907) and Raven (1950) stated that in their gorillas the lumbricales I and II were innervated by the median nerve, and the lumbricales III and IV by the deep branch of ulnar nerve. Preuschoft (1965) stated that he found this latter condition in one of the three specimens dissected by him, but that in the two other specimens that he dissected the median nerve supplied the lumbricalis I only.
- Notes: As reported by Raven (1950) and Gibbs (1999), and found in our VUG specimen, the lumbricalis IV was considerably smaller than the other lumbricales (see weights above). The **contrahentes digitorum** were not present as distinct, fleshy muscles in the gorilla specimens dissected by Macalister (1873), Deniker

(1885), Hartmann (1886), Hepburn (1892), Sommer (1907), Pira (1913), Raven (1950), Preuschoft (1965) and Sarmiento (1994) and by us. In the review of the literature by Sarmiento they were missing in 7 out of 7 gorillas, but Day and Napier (1963) stated that they did find one contrahens in one of the two gorillas dissected by them. The **intercapitulares** (see, e.g., Jouffroy 1971) are also usually not present in gorillas.

Adductor pollicis (15.0 g; Figs. 37a,b, 38a,b, 40, 41, 43, 44, 46, 47, 48)

- Usual attachments: Caput obliquum and caput transversum are usually well differentiated, connecting the metacarpals I, II and/or III, the contrahens fascia, and often at least some carpal bones and/or ligaments, to the metacarpophalangeal joint, as well as to the proximal phalanx and (by means of a thin tendon) also to the distal phalanx of the thumb (see Figs. 46, 47); when present, the TDAS-AD (see notes below) often connects the proximo-medial portion of the metacarpal I and/or adjacent carpal structures to the metacarpophalangeal joint and/or the proximal portion of the proximal phalanx of digit 1 (see Fig. 47).

- Usual innervation: Deep branch of the ulnar nerve (Eisler 1890, Hepburn 1892, Sommer 1907, Preuschoft 1965). Raven (1950) stated that in the gorilla dissected by him, the adductor pollicis had a transverse head (called in that work 'adductor pollicis') innervated by the ulnar nerve and an oblique head (called in that work as 'part of the inner head of the flexor pollicis brevis') innervated by the 'n. medianus'. However, because there is some confusion in that work, between the deep branch of the ulnar nerve (often called 'ulnar branch of medianus n.' in that work) and the median nerve, and also because it was not Raven, but J.E. Hill, who wrote in that work, that the oblique head was innervated by the 'n. medianus', it is not clear if the oblique head of that gorilla was really innervated by the median nerve, or was, instead, innervated by the ulnar nerve, as is often the case in other primates (and in the gorillas described by other authors, such as Eisler 1890, Hepburn 1892, Sommer 1907 and Preuschoft 1965).

- Notes: There has been a lot of controversy regarding the homologies of the thenar muscles of primate and non-primate mammals. This subject has been discussed in detail in the recent works of Diogo et al. (2009a) and in particular of Diogo and Abdala (2010), and here we will only summarize the main conclusions of those works. The '**interosseous volaris primus of Henle**' of modern human anatomy corresponds most likely to a **thin, deep additional slip of the adductor pollicis** (**TDAS-AD** *sensu* Diogo and Abdala 2010), and not to the **flexor brevis profundus 2** of 'lower' mammals, as suggested by some authors (and by the erroneous use of the name 'interosseous volaris primus of Henle'). Among gorillas, this structure was described by Huxley (1864) and Brooks (1886a) and was also present in our CMS GG1 specimen (see Fig. 47) and in one of the four specimens dissected by Preuschoft (1965), but is absent in most gorillas according to other authors . For instance, it was absent in 2 out of 2 gorillas dissected by Sarmiento (1994) and in 10 out of 11 gorillas reported in the literature reviewed by him, and was also absent in the gorilla dissected by Susman et al. (1999) and in 3 of the 4 gorillas dissected by Preuschoft (1965), as well as in our VU GG1 specimen. Regarding the '**deep head of the flexor pollicis brevis**' of modern human anatomy, this corresponds very likely to the flexor brevis profundus 2 of 'lower' mammals. We found this structure in the gorillas dissected by us, and it was also present in the specimens reported by Hepburn (1892), Huxley (1864), Macalister (1873), Bischoff (1880) and Hartmann (1886), although it is possible, and even likely, that part or the totality of the structure that some of these authors called 'deep head of the flexor pollicis brevis' actually corresponds to part or the totality of the oblique head of the adductor pollicis *sensu* the present work. Regarding the '**superficial head of the flexor pollicis brevis**' of modern human anatomy, this seems to correspond to a true **flexor pollicis brevis**. This true flexor pollicis brevis and the **opponens pollicis** derive very likely from the **flexor brevis profundus 1** of 'lower' mammals, and are present as distinct muscles in the vast majority of gorillas. Regarding the oblique and transverse heads of the adductor pollicis, in the specimens shown in Fig. 24 of Owen (1868) and Fig. 3 of Macalister (1873) these structures seem to have a partial insertion on the distal phalanx of the thumb, and such a partial insertion was also found in the specimens dissected by Duvernoy (1855-1856), Duckworth (1904), Raven (1950) and Preuschoft (1965) and in our CMS GG1 and VU GG1 specimens; this configuration is commonly present in gorillas according to Tuttle (1969) and Marzke (1997), while Sommer (1907) only described an insertion on the proximal phalanx of digit 1 in the specimen dissected by him. A partial insertion of the transverse and/or oblique heads of the adductor pollicis onto a small portion of metacarpal I was described in the gorillas dissected by Raven (1950) and Preuschoft (1965). Jouffroy and Lessertisseur (1960) suggested that such an insertion on the metacarpal is often present in gorillas, but in the vast majority of the gorillas described and/or illustrated in the literature, as well as in the gorillas dissected by us, there is actually no insertion on this bone.

- Synonymy: Adductor pollicis and possibly also part or the totality of the inner head of the flexor pollicis brevis (Macalister 1873, Raven 1950, Preuschoft 1965); contrahentes I (Jouffroy 1971); see notes above.

Interossei palmares (I: 4.6 g; II: 6.7 g; III: 5.4 g; Figs. 46, 48, 49)
- Usual attachments: Interosseous palmaris I mainly from metacarpal II to the ulnar side of the proximal phalanx and of the extensor expansion of digit 2; interosseous palmaris II mainly from metacarpal IV to the radial side of the proximal phalanx and of the extensor expansion of digit 4; interosseous palmaris III mainly from metacarpal V to the radial side of the proximal phalanx and of the extensor expansion of digit 5.

- Usual innervation: Deep branch of the ulnar nerve (Eisler 1890, Hepburn 1892, Sommer 1907, Raven 1950, Preuschoft 1965).
- Notes: See notes on interossei dorsales.
- Synonymy: Part of adductors (Macalister 1873); part of palmar interossei (Preuschoft 1965).

Interossei dorsales (I: 24.1 g; II: 13.5 g; III: 9.4 g; IV: 10.0 g; Figs. 31, 37a,b, 38a,b, 40, 41, 43, 44, 46, 48, 49, 50)

- Usual attachments: Interosseous dorsalis I mainly from metacarpals I and II to the radial side of the proximal phalanx and of the extensor expansion of digit 2; interosseous dorsalis II mainly from metacarpals II and III to the radial side of the proximal phalanx and of the extensor expansion of digit 3; interosseous dorsalis III mainly from metacarpals III and IV to the ulnar side of the proximal phalanx and of the extensor expansion of digit 3; interosseous dorsalis IV mainly from metacarpals IV and V to the ulnar side of the proximal phalanx and of the extensor expansion of digit 4.
- Usual innervation: Deep branch of the ulnar nerve (Eisler 1890, Hepburn 1892, Sommer 1907, Raven 1950, Preuschoft 1965).
- Notes: As explained above (see adductor pollicis), the **flexor brevis profundus 2** of 'lower' mammals corresponds very likely to the '**deep head of the flexor pollicis brevis**' of modern human anatomy; the '**superficial head of the flexor pollicis brevis**' of modern human anatomy, as well as the **opponens pollicis**, derive very likely from the **flexor brevis profundus 1** of 'lower' mammals, while the **flexor digiti minimi brevis** and the **opponens digiti minimi** derive very likely from the **flexor brevis profundus 10** of 'lower mammals' (for a recent review, see Diogo et al. 2009a and Diogo and Abdala 2010). Contrary to *Pan*, in *Gorilla* and other hominoids, including modern humans, the **flexores breves profundi** 3, 5, 6 and 8 are fused with the **intermetacarpales** 1, 2, 3 and 4, forming the interossei dorsales 1, 2, 3 and 4, respectively; the **interossei palmares** 1, 2 and 3 of gorillas, modern humans, hylobatids and orangutans thus correspond respectively to the flexores breves profundi 4, 7 and 9 of 'lower' mammals (see, e.g., Lewis 1989, Diogo et al. 2009a, and Diogo and Abdala 2010). However, in at least some of the interossei dorsales of the gorillas dissected by us it is still possible to identify the portion derived from the intermetacarpales and the portion derived from the flexores breves profundi. This is illustrated in Fig. 49, which shows the interosseous dorsalis I of the CMS GG1 specimen: the most ventral portion of this muscle corresponds to the flexor brevis profundus 3 of 'lower' mammals, originating mainly from the ventral margin of metacarpal II and forming a tendon that passes superficially to the transverse lamina of digit 2; the most dorsal portion of the muscle corresponds to the intermetacarpales 3 of 'lower' mammals, originating mainly from the medial (ulnar) margin of metacarpal I and the lateral (radial) margin of metacarpal II and forming a

tendon that goes, or passes deep, to the transverse lamina of digit 2. A similar configuration seems to be found in the specimens dissected by other authors. For instance, Preuschoft (1965) described four 'dorsal interossei' and seven 'palmar interossei' in the gorillas dissected by him. The former structures clearly seem to correspond to the intermetacarpales 1, 2, 3 and 4 of 'lower' mammals, while the latter structures clearly seem to correspond to the flexores breves profundi 3, 4, 5, 6, 7, 8 and 9 of 'lower' mammals. This indicates that in those gorillas the interossei dorsales 1, 2, 3 and 4 *sensu* the present work can still be differentiated into the portions derived from the flexores breves profundi 3, 5, 6 and 8 and the portions derived from the intermetacarpales 1, 2, 3 and 4, respectively. That is, three of the 'palmar interossei' *sensu* Preuschoft (1965) actually correspond to the interossei palmares *sensu* the present work, while the other four 'palmar interossei' and the four 'dorsal interossei' *sensu* Preuschoft (1965) actually form the interossei dorsales *sensu* the present work. The **interdigitales**, present in primates such as lorisoids, and the **interossei accessorii**, present in primates such as hylobatids and some strepsirhines (see, e.g., Jouffroy 1971), are usually not present in gorillas.

- Synonymy: Abductors and extra adductor of digit 3 (Macalister 1873); dorsal interossei plus part of palmar interossei (Preuschoft 1965).

Flexor pollicis brevis ('deep head' plus 'superficial head': 4.7 g; Figs. 41, 43, 44, 48)

- Usual attachments: As explained above (see adductor pollicis), the **flexor brevis profundus 2** of 'lower' mammals corresponds very likely to the '**deep head of the flexor pollicis brevis**' of modern human anatomy; the '**superficial head of the flexor pollicis brevis**' of modern human anatomy thus corresponds to a true flexor pollicis brevis, being most likely derived, together with the **opponens pollicis**, from the **flexor brevis profundus 1** of 'lower' mammals (for a recent review, see Diogo et al. 2009a and Diogo and Abadala 2010). As there is some confusion in the literature about the so-called 'superficial and deep heads of the flexor pollicis brevis' in gorillas (and other primates), we will describe the condition found in our VU GG1 specimen, which is illustrated in Figs. 41, 43, 44 and 48. In both sides of this specimen, there was a distinct 'deep head' (flexor brevis profundus 2 *sensu* Diogo and Abdala 2010), which was partially blended with the opponens pollicis and mainly connected the flexor retinaculum to the metacarpophalangeal joint and the base of the proximal phalanx of digit 1. The 'superficial head' (true flexor pollicis brevis *sensu* Diogo and Abdala 2010, derived, together with the opponens pollicis, from the flexor brevis profundus 2) originated from the flexor retinaculum and inserted onto the radial side of the metacarpophalangeal joint and of the base of the proximal phalanx of digit 1.

- Usual innervation: In gorillas, the so-called 'superficial head' is usually innervated by the median nerve (Brooks 1887, Eisler 1890, Hepburn 1892,

Sommer 1907, Raven 1950, Preuschoft 1965). However, the innervation of the so-called 'deep head' is less clear. For instance, as explained above (see adductor pollicis), in the work of Raven (1950) it is stated that the 'inner head of the flexor pollicis brevis' was innervated by the 'n. medianus' . However, as there is some confusion in that work between this structure and the oblique head of the adductor pollicis, as well as between the deep branch of the ulnar nerve (often called as 'ulnar branch of medianus n.' in that work) and the median nerve, and also because it was actually not Raven, but J.E. Hill, who wrote in that work that the 'inner head of the flexor pollicis brevis' was innervated by the 'n. medianus', it is not clear if the so-called 'deep head' of gorillas is really innervated by the median nerve, or is, instead, innervated by the ulnar nerve or by both the median and ulnar nerves, as is often the case in other primates, including modern humans (see, e.g., Day and Napier 1963, Swindler and Wood 1973, Lewis 1989).
* Notes: See notes about adductor pollicis.

Opponens pollicis (6.4 g; Figs. 41, 43, 48)
* Usual attachments: From the flexor retinaculum, trapezium, and, often, the adjacent sesamoid bone, to the whole length of metacarpal I.
* Usual innervation: Median nerve (Hepburn 1892, Sommer 1907, Raven 1950, Preuschoft 1965).

Flexor digiti minimi brevis (8.1 g; Figs. 39, 41, 42, 46, 48)
* Usual attachments: From the hamate and/or flexor retinaculum and, occasionally, from the pisiform, to the ulnar side of the proximal phalanx and, often, of metacarpophalangeal joint and the extensor expansion of digit 5.
* Usual innervation: Deep branch of the ulnar nerve (Eisler 1890, Hepburn 1892, Höfer 1892, Sommer 1907, Raven 1950, Preuschoft 1965).
* Notes: Macalister (1873), Duckworth (1904) and Raven (1950) described an origin of this muscle from the hamate, and Preuschoft (1965) also found an origin from the hamate in two of his four specimens. However, in the two other specimens dissected by Preuschoft (1965) there was a direct origin from the pisiform. In our VU GG1 and CMS GG1 specimens the origin was from the hamate and flexor retinaculum, as described by Hepburn (1892). Most authors described an insertion on the ulnar side of the proximal phalanx of digit 5, but Raven (1950) reported an insertion on the proximal phalanx, the metacarpophalangeal joint, and the extensor expansion of this digit. In our VU GG1 specimen the insertion seemed to be similar to that reported by Raven (1950).
* Synonymy: Flexor digiti quinti (Sommer 1907); flexor digiti quinti brevis (Raven 1950, Jouffroy 1971); flexor brevis digiti V (Preuschoft 1965).

Opponens digiti minimi (7.0 g; Figs. 37, 38, 42, 46, 48)
* Usual attachments: From the hamate and, often, from the flexor retinaculum, to the whole length of metacarpal V.

- Usual innervation: Deep branch of the ulnar nerve (Eisler 1890, Hepburn 1892, Höfer 1892, Sommer 1907, Raven 1950, Preuschoft 1965).
- Notes: Lewis (1989) stated that the opponens digiti minimi is more markedly divided in hominoids such as *Pan* and in modern humans than in hominoids such as hylobatids, and Brooks (1886a) stated that, contrary to the condition in *Pan* and modern humans, in hominoids such as *Pongo* there are no superficial and deep bundles of the muscle separated by the deep branch of the ulnar nerve. In both hands of our CMS GG1 specimen the opponens digiti minimi was divided into deep and superficial heads, the deep branch of the ulnar nerve passing mainly radial to these heads, and only slightly ventral to the deep head and dorsal to the superficial head. In our VU GG1 specimen the deep branch of the ulnar nerve passed mainly radial to the opponens digiti minimi, but there was a very small, distinct portion of this muscle, of about 5 square mm, going from the hamate to a proximo-radial portion of metacarpal V, which did seem to pass dorsally to the deep branch of the ulnar nerve. It is not clear if this was an additional bundle of the opponens pollicis, or if it was a poorly differentiated deep head of the opponens digiti minimi *sensu* Lewis (1989).
- Synonymy: Opponens digiti quinti (Sommer 1907, Raven 1950); opponens digiti V (Preuschoft 1965).

Abductor pollicis brevis (7.4 g; Figs. 41, 46)

- Usual attachments: From the flexor retinaculum, the trapezium, and, at least in some specimens, the adjacent sesamoid bone, to the radial side of the metacarpophalangeal joint and of the base of the proximal phalanx of digit 1 and, occasionally, also to the distal phalanx of this digit and/or to the distal portion of metacarpal I.
- Usual innervation: Median nerve (Brooks 1887, Eisler 1890, Hepburn 1892, Sommer 1907, Raven 1950, Preuschoft 1965).
- Notes: On one side of our VU GG1 specimen, the abductor pollicis brevis was partially inserted onto the distal phalanx of the thumb; Raven (1950) and Preuschoft (1965) stated that, in the gorillas dissected by them, there was a partial insertion onto the distal portion of metacarpal I.

Abductor digiti minimi (12.7 g; Figs. 39, 41, 46, 48)

- Usual attachments: From the pisiform, and occasionally from the adjacent ligaments, as well as from the metacarpal V, the hamate and/or the triquetrum, to the ulnar side of the metacarpophalangeal joint and the base of the proximal phalanx of digit 5.
- Usual innervation: Deep branch of the ulnar nerve (Eisler 1890, Hepburn 1892, Höfer 1892, Sommer 1907, Raven 1950, Preuschoft 1965).
- Notes: Duckworth (1904) described a gorilla in which this muscle originated from the pisiform, the adjacent ligaments, and the metacarpal V, while Preuschoft (1965) stated that in the four gorillas he dissected the insertion was from the flexor

retinaculum, hamate and pisiform and, in one of the four specimens, also from the triquetrum.

- Synonymy: Abductor digiti quinti (Sommer 1907, Raven 1950); abductor digiti V (Preuschoft 1965); abductor digiti quinti brevis (Jouffroy 1971).

Extensor carpi radialis longus (65.9 g; Figs. 24, 30, 31, 35)

- Usual attachments: From the lateral supracondylar ridge and/or the lateral epicondyle of the humerus to the base of the metacarpal II.
- Usual innervation: Radial nerve (Eisler 1890, Hepburn 1892, Höfer 1892, Duckworth 1904, Sommer 1907, Raven 1950, Preuschoft 1965).
- Notes: The statements of Bojsen-Moller (1978) are confusing and seem to contradict the vast majority of the descriptions provided in the literature and also our own observations (according to which the muscle inserts exclusively onto metacarpal II), as he states that in 2 of the 4 gorillas dissected by him the muscle goes to metacarpals II and I. This is very likely related to the fact that he considered the intermetacarpal ligament connecting these two metacarpals to be part of the extensor carpi radialis longus.
- Synonymy: Extensor carpi radialis longior (Chapman 1878, Hepburn 1892); premier radial (Deniker 1885).

Extensor carpi radialis brevis (59.1 g; Figs. 30, 31, 35)

- Usual attachments: From the lateral epicondyle, and occasionally also from the lateral condyle of the humerus, to the base of metacarpal III and, often, also to base of metacarpal II
- Usual innervation: Radial nerve (Eisler 1890, Duckworth 1904, Sommer 1907, Raven 1950, Preuschoft 1965); posterior interosseous nerve (Hepburn 1892).
- Notes: Deniker (1885) reported an origin from both the lateral condyle and the lateral epicondyle of the humerus, while in the specimen dissected by Raven (1950) and also in our VU GG1 specimen, there was an insertion on both metacarpals II and III.
- Synonymy: Extensor carpi radialis brevior (Chapman 1878, Hepburn 1892); deuxième radial (Deniker 1885).

Brachioradialis (201.9 g; Figs. 19, 23, 25, 26, 27, 28, 29, 30, 37a,b, 38a,b)

- Usual attachments: From the humeral shaft (at least sometimes extending proximally to the insertion of the deltoideus), to the radius (usually extending distally to the styloid process).
- Usual innervation: Radial nerve (Eisler 1890, Hepburn 1892, Höfer 1892, Duckworth 1904, Sommer 1907, Straus 1941a,b, Raven 1950, Preuschoft 1965).
- Notes: In our VU GG1 specimen the brachioradialis ran from the distal 12.5 cm of the humerus (of the total 36 cm of this bone) to the distal portion of the radius, including the styloid process.

- Synonymy: Supinator longus (Owen 1868, Macalister 1873, Barnard 1875, Deniker 1885, Hartmann 1886); supinator radii longus (Hepburn 1892).

Supinator (79.5 g; Figs. 23, 24, 25, 26, 27, 28, 29, 34)
- Usual attachments: From the lateral epicondyle of the humerus (caput humerale, or superficiale) and the proximal portion of the ulna (caput ulnare, or profundum) to the proximal portion of the radius.
- Usual innervation: Radial nerve (Eisler 1890, Duckworth 1904, Sommer 1907, Raven 1950, Preuschoft 1965); posterior interosseous nerve (Hepburn 1892).
- Notes: An origin from both the ulna and the lateral epicondyle of the humerus was found in the specimens reported in Raven (1950), Preuschoft (1965), as well as in our VU GG1 specimen. The supinator was pierced by the radial nerve in, e.g., three gorillas dissected by Preuschoft (1965) and on one side of a gorilla examined by Eisler (1890), although on the other side of this latter gorilla the nerve apparently did not pierce the muscle. In our VU GG1 specimen the nerve seemed to pierce the muscle.
- Synonymy: Supinator brevis (Chapman 1878, Hepburn 1892).

Extensor carpi ulnaris (51.7 g; Figs. 30, 31, 32, 33, 35)
- Usual attachments: From the lateral epicondyle of the humerus (caput humerale) and the ulna (caput ulnare) to metacarpal V.
- Usual innervation: Radial nerve (Eisler 1890, Sommer 1907, Raven 1950, Preuschoft 1965); posterior interosseous nerve (Hepburn 1892, Duckworth 1904).
- Notes: An origin from the humerus and ulna was found in, e.g., the specimen dissected by Raven (1950) and our VU GG1 specimen, while an origin from the ulna and indirectly from the humerus—via the antebrachial fascia, intermuscular septum and/or the tendon of the extensor digitorum—was found in the specimens dissected by Preuschoft (1965).
- Synonymy: Cubital postérieur (Deniker 1885).

Anconeus (8.8 g; Figs. 30, 32, 33)
- Usual attachments: From the lateral epicondyle of the humerus to the olecranon process and, at least in some specimens, to the region of the ulna lying just distal to the olecranon process.
- Usual innervation: Radial nerve (Eisler 1890, Hepburn 1892, Sommer 1907, Raven 1950, Preuschoft 1965).

Extensor digitorum (96.2 g; Figs. 30, 31, 32, 33, 35, 37a,b)
- Usual attachments: From the lateral epicondyle of the humerus and, occasionally, also from the radius and/or ulna, to the middle phalanges and (via the extensor expansions) also to the distal phalanges of digits 2, 3, 4 and 5, although in a few cases there is no insertion on digit 5.
- Usual innervation: Radial nerve (Eisler 1890, Höfer 1892, Sommer 1907, Straus 1941a,b, Raven 1950, Preuschoft 1965); posterior interosseous nerve (Hepburn 1892, Duckworth 1904).

- Notes: In one gorilla described by Raven (1950) and three gorillas reported by Preuschoft 1965), as well as in our VU GG1 specimen, the bony origin was exclusively from the lateral epicondyle of the humerus, while in one specimen reported by Deniker (1885) and one specimen described by Duckworth (1904) the bony origin was from the lateral epicondyle and from the ulna and radius. The insertion of the muscle is usually on digits 2–5, as described by Owen (1868), Macalister (1873), Chapman (1878), Hepburn (1892), Sommer (1907), Raven (1950), Preuschoft (1965) and Kaneff (1979) and corroborated by our dissections, but there was an insertion onto digits 2–4 in one specimen dissected by Deniker (1885) and on the right side of one specimen reported by Preuschoft (1965).
- Synonymy: Common extensor of the fingers (Owen 1868, Deniker 1885); extensor digitorum communis (Barnard 1875, Sommer 1907, Straus 1941a, b, Raven 1950, Preuschoft 1965, Jouffroy 1971); extensor communis digitorum (Chapman 1878, Hepburn 1892).

Extensor digiti minimi (14.7 g; Figs. 30, 31, 32, 33, 35, 37)
- Usual attachments: From the lateral epicondyle of the humerus and/or common extensor tendon, to the middle phalanx and (via the extensor expansion) also to the distal phalanx of digit 5, although in a few cases there is an insertion on digit 4.
- Usual innervation: Radial nerve (Eisler 1890, Höfer 1892, Sommer 1907, Straus 1941a,b, Raven 1950, Preuschoft 1965); posterior interosseous nerve (Hepburn 1892, Duckworth 1904).
- Notes: In the gorillas described by Owen (1868), Chapman (1878), Bischoff (1880), Deniker (1885), Hepburn (1892), Sommer (1907), Pira (1913), Straus (1941a), Raven (1950), Preuschoft (1965), Kaneff (1980a) and Aziz and Dunlap (1986), and in those dissected by us, the muscle goes to digit 5 only. However, an insertion on digits 4 and 5 was reported on one side of one specimen dissected by Preuschoft (1965), and an insertion on digit 4 only was found in the specimen dissected by Macalister (1873). According to the literature review done by Gibbs (1999) an insertion to digit 4 occurs in only 1 out of 14 gorillas, and according to the literature review done by Straus (1941a) this occurs in only 7% of gorillas. The **extensor digiti quarti** is usually not present as a distinct muscle in gorillas.
- Synonymy: Extensor minimi digiti (Owen 1868, Chapman 1878, Hepburn 1892); extenseur propre du petit doigt (Deniker 1885); extensor digiti quinti proprius (Sommer 1907, Pira 1913, Raven 1950); extensor digiti-quarti et-quinti proprius (Straus 1941a,b); extensor digiti V (Preuschoft 1965); extensor lateralis (Jouffroy 1971); extensor digitorum proprius or profundus 5 (Lewis 1989).

Extensor indicis (6.6 g; Figs. 31, 34, 35, 36, 37)
- Usual attachments: From the radius, ulna and/or interosseous membrane to digit 2, although occasionally there is also, or only, an insertion on metacarpals III and/or IV.
- Usual innervation: Posterior interosseous nerve (Eisler 1890, Höfer 1892, Sommer 1907, Straus 1941a,b, Raven 1950, Preuschoft 1965).
- Notes: Among gorillas, a bony origin from the radius was reported by Deniker (1885), and found in our VU GG1 specimen; a bony origin from the ulna was described by Raven (1950) and Preuschoft (1965). An insertion of the extensor indicis onto digit 2 was found by most authors, including Owen (1868), Macalister (1873), Barnard (1875), Chapman (1878), Deniker (1885), Hepburn (1892), Sommer (1907), Straus (1941a), Preuschoft (1965), Aziz and Dunlap (1986) and Lewis (1989), and it was also found in our CMS GG1 specimen (see Fig. 37b). According to the literature review by Straus (1941a,b), an insertion on digit 2 occurs in about 100% of gorillas. However, Raven (1950) described an insertion onto digit 2 and onto the hamate and capitate. Kaneff (1980a,b) reported an insertion onto metacarpals III and/or IV only: this configuration was found in one side of our VU GG1 specimen (see Fig. 31), the other side showing an attachment to these two metacarpals and to digit 2 (see Fig. 36). The **extensor communis pollicis et indicis**, the **extensor digiti III proprius** and the **extensor brevis digitorum manus** (see, Diogo et al. 2009a) are usually not present as distinct muscles in gorillas. However, in his Fig. 8.4A, Lewis (1989) did show a gorilla in which the extensor indicis gives rise to a tendon to digit 2 and to a small muscle extensor brevis digitorum manus, which then gives rise to a tendon to digit 3.
- Synonymy: Indicator (Owen 1868); part, or all, of the extensor profundus digitorum or extensor digitorum profundus (Barnard 1875, Hepburn 1892, Straus 1941a); extenseur propre de l'index (Deniker 1885); extensor indicis proprius (Sommer 1907, Raven 1950); extensor digiti II proprius or extensor digitorum profundus proprius (Aziz and Dunlap 1986); extensor digitorum proprius or profundus 2 (Lewis 1989).

Extensor pollicis longus (13.2 g; Figs. 31, 33, 34, 35, 36, 37)
- Usual attachments: From the ulna, interosseous membrane and/or radius, to the distal phalanx of the thumb, although occasionally there is also an insertion on the proximal phalanx and/or the metacarpophalangeal joint of this digit, and/or to digit 2.
- Usual innervation: Posterior interosseous nerve (Eisler 1890, Hepburn 1892, Höfer 1892, Duckworth 1904, Sommer 1907, Straus 1941a,b, Raven 1950, Preuschoft 1965).

- Notes: Among gorillas, a bony origin from the radius was found in our VU GG1 specimen; a bony origin from the ulna was described by Raven (1950) and Preuschoft (1965). According to Hepburn (1892), the extensor pollicis longus of the gorilla dissected by him sent a tendon to the distal phalanx of the thumb, as well as a tendon to digit 2 and an extra, small tendon to the thumb. Raven (1950) stated that in the gorilla dissected by him this muscle was attached to the proximal and distal phalanges of the thumb. Straus (1941a,b) described a partial attachment onto the metacarpophalangeal joint of the thumb.
- Synonymy: Extensor of the last phalanx of the pollex (Owen 1868); extensor secundi internodii pollicis (Chapman 1878, Hepburn 1892); part or totality of extensor profundus digitorum or extensor digitorum profundus (Straus 1941a,b); extensor digitorum proprius or profundus 1 (Lewis 1989).

Abductor pollicis longus (82.1 g; Figs. 31, 32, 33, 34, 35, 36, 38a,b)
- Usual attachments: From the radius, interosseous membrane and ulna to metacarpal I, trapezium and/or adjacent to the sesamoid bone, and, often, to the proximal phalanx of digit 1.
- Usual innervation: Radial nerve (Eisler 1890, Hepburn 1892, Höfer 1892, Sommer 1907, Straus 1941a,b, Raven 1950, Preuschoft 1965); posterior interosseous nerve (Duckworth 1904).
- Notes: It should be noted that some authors described an '**extensor pollicis brevis**' and an 'abductor pollicis longus' in gorillas (e.g., Hepburn 1892, Straus 1941a,b, Raven 1950, Preuschoft 1965, Sarmiento 1994), because in these primates a tendon of the abductor pollicis longus (*sensu* the present work) often inserts onto the proximal phalanx of the thumb, i.e., to the typical insertion point of the extensor pollicis brevis of modern humans. However, as stressed by authors such as Huxley (1864), Macalister (1873), Bischoff (1880), Deniker (1885), Tuttle (1970), Kaneff (1979, 1980a,b) and Aziz and Dunlap (1986), and as corroborated by our dissections (see, e.g., Fig. 31), in gorillas there is usually a single fleshy belly of the abductor pollicis longus that then gives rise to the so-called 'tendons of the extensor pollicis brevis and of the abductor pollicis longus', as is usually the case in other primates, other than hylobatids and modern humans. That is, contrary to modern humans and hylobatids, in *Pongo, Pan, Gorilla* and in most, if not all, other primates the extensor pollicis brevis is effectively usually not present as a separate, distinct, muscle. Among gorillas, the attachment of the abductor pollicis longus is onto the metacarpal I and the proximal phalanx according to Owen (1868), Chapman (1878), Bischoff (1880), Hepburn (1892), Pira (1913), Straus (1941a,b), and to our dissection of CMS GG1 (see Fig. 38b); to the carpal/metacarpal region and the proximal phalanx according to Duckworth (1904), Kaneff (1980a,b) and Tuttle (1969); to the trapezium, metacarpal I and proximal phalanx in the specimen described by Raven (1950), in one specimen plus one side of another specimen of Preuschoft (1965); to the

trapezium, metacarpal I and metacarpophalangeal joint of the thumb on one side of one specimen plus in one other specimen of Preuschoft (1965); to the metacarpal I and trapezium/sesamoid bone according to Macalister (1873), Hartmann (1886), and to our dissection of VU GG1 (see, e.g., Fig. 31); to the metacarpal I and the carpal region in the gorilla fetus of Deniker (1885). An insertion onto the proximal phalanx of digit 1, trapezium and metacarpal I is the modal condition in *Gorilla* according to Gibbs (1999), and in the literature review done by Keith (1899) this condition was found in 4 out of 9 gorillas. Sarmiento (1994) found an extension to the proximal phalanx in 1 of the 3 gorillas dissected by him, but in the review of the literature done by him, he found that there was such an extension in 9 out of 17 gorillas, and according to the literature review done by Straus (1941a,b) an insertion to the proximal phalanx of the thumb occurs in 53% of gorillas.

• Synonymy: Extensor of the metacarpal of the pollex plus extensor of the first phalanx of the pollex (Owen 1868); extensor ossis metacarpi pollicis plus extensor primi internodii pollicis (Chapman 1878, Hepburn 1892); extensor pollicis brevis plus abductor pollicis longus (Raven 1950, Preuschoft 1965, Sarmiento 1994).

CHAPTER 5

Trunk and Back Musculature

Obliquus capitis inferior (Fig. 51)
- Usual attachments: From the spinous process of C2 to the transverse process of C1.
- Usual innervation: Branch of dorsal ramus of C1 (Raven 1950, Gibbs 1999).

Obliquus capitis superior (Fig. 51)
- Usual attachments: From the transverse process of C1 to the occipital bone.
- Usual innervation: Branch of dorsal ramus of C1 (Raven 1950, Gibbs 1999).

Rectus capitis anterior
- Usual attachments: From the body of the atlas to the occipital bone.
- Usual innervation: Branch of ventral ramus of C1 (Raven 1950, Gibbs 1999).

Rectus capitis lateralis
- Usual attachments: From the transverse process of the atlas to the occipital bone.
- Usual innervation: Branch of ventral ramus of C1 (Raven 1950, Gibbs 1999).

Rectus capitis posterior major (Fig. 51)
- Usual attachments: From the spinous process of C2 to the occipital bone.
- Usual innervation: Branch of dorsal ramus of C1 (Raven 1950, Gibbs 1999).
- Notes: In gorillas this muscle is usually smaller than the rectus capitis posterior minor, according to the literature review of Gibbs (1999).

Rectus capitis posterior minor (Fig. 51)
- Usual attachments: From the posterior tubercle of C1 to the region of inferior nuchal line.
- Usual innervation: Branch of dorsal ramus of C1 (Raven 1950, Gibbs 1999).

Longus capitis
- Usual attachments: From the transverse processes of C2–C6 to the basiocciput.
- Usual innervation: Branches of ventral rami of C1–C5 (Raven 1950, Gibbs 1999).

Longus colli
- Usual attachments: From the ventral tubercles of C3–C6 and bodies of C2–C6, to the transverse processes of C2 and C1, ventral tubercle of C1, and the transverse process of C6.
- Usual innervation: Branches of ventral rami of C2–C7 (Raven 1950, Gibbs 1999).

Scalenus anterior (Fig. 6)
- Usual attachments: From C5 and C6 to scalene tubercle of rib 1 (e.g., Raven 1950, Gibbs 1999); Stewart (1936) reported an origin from C3–C6.
- Usual innervation: C5–C7 (Raven 1950, Gibbs 1999).

Scalenus medius
- Usual attachments: From C6 and C7 to rib 1 (e.g., Raven 1950, Gibbs 1999); but see notes below.
- Usual innervation: C7–T1 (Raven 1950, Gibbs 1999).
- Notes: According to Deniker (1885) the scalenus medius is completely absent in the fetal and juvenile gorilla specimens he dissected. Stewart (1936) stated that the scalenus medius (scalenus medius plus scalenus posterior *sensu* the present work) runs from the transverse processes of the second and third cervical vertebrae to the first rib; an extra slip, which according to him might correspond to a '**scalenus minimus**', runs from the transverse processes of C6 and C7 to the first rib. Jouffroy (1971) stated that authors such as Raven (1950) describe a scalenus posterior in gorillas, but that this is probably a continuation of the scalenus medius. Jouffroy (1971) suggests that these two muscles are usually fused in gorillas, as they usually are in *Pan* and *Pongo*.

Scalenus posterior
- Usual attachments: From C2–C4 to ribs 1 and 2 (e.g., Raven 1950, Gibbs 1999); but see the notes about scalenus medius.
- Usual innervation: C3–C7 (Raven 1950, Gibbs 1999).
- Notes: See notes about scalenus medius.

Levatores costarum
- Usual attachments: From C7–T12 to the ribs lying posteriorly to these vertebrae.
- Usual innervation: Data not available.

Intercostales externi (Fig. 55)
- Usual attachments: Connect the adjacent margins of each pair of ipsilateral ribs.
- Usual innervation: Intercostal nerves T2–T13 (Raven 1950, Gibbs 1999).

Intercostales interni (Fig. 55)
- Usual attachments: Connect the adjacent margins of each pair of ipsilateral ribs.
- Usual innervation: Intercostal nerves T2–T13 (Raven 1950, Gibbs 1999).

Transversus thoracis
- Usual attachments: From the xiphoid process and the body of the sternum to the internal surface of ipsilateral ribs 1–8.
- Usual innervation: Intercostal nerves T2–T6 (Raven 1950, Gibbs 1999).
- Notes: According to Raven (1950), the **subcostales** were apparently absent in the gorilla dissected by him; he did also not report **intercostales intimi** in this specimen.

Splenius capitis (Figs. 6, 21, 22, 51)
- Usual attachments: From C3–C7 to the nuchal crest and the mastoid process.
- Usual innervation: Data not available.
- Synonymy: Part of splenius cervicus et capitis (Raven 1950).

Splenius cervicis
- Usual attachments: From T1–T3 to the transverse processes of C1 and C2.
- Usual innervation: Data not available.
- Synonymy: Part of splenius cervicus et capitis (Raven 1950).

Serratus posterior superior (Fig. 52)
- Usual attachments: From C1–C7 to ribs 2–6.
- Usual innervation: Intercostal nerves 1–5 (Raven 1950, Gibbs 1999).
- Notes: According to Raven (1950), the **serratus posterior inferior** was absent in the gorilla dissected by him; this latter muscle also seemed to be absent in our VU GG1 specimen.

Iliocostalis (Fig. 52)
- Usual attachments: From the iliac crest, thoracolumbar fascia and all ribs except the first two, to ribs and C2–C4.
- Usual innervation: Data not available.

Longissimus (Fig. 52)
- Usual attachments: From C5–T8 to the ventral tubercles of C2–C7 (pars cervicis); from C2–T1 to the mastoid process and atlas (pars capitis); from the iliac crest, thoracolumbar fascia and sacrum to lumbar fascia, lumbar transverse processes and ribs (pars thoracis).
- Usual innervation: According to Raven (1950), the pars capitis is innervated by C2–C8; data are not available for the other parts of the muscle.
- Notes: As in modern humans, in gorillas the **atlantomastoideus** might be occasionally present as a distinct muscle, running from the atlas to the mastoid process (see Sommer 1907).
- Synonymy: Longissimus atlantis et capitis plus pars cervicis of longissimus plus longissimus dorsi (Raven 1950).

Spinalis (Figs. 51, 52)
- Usual attachments: From the spinous processes and supraspinous ligament of T9–T13 to the semispinalis thoracis, semispinalis cervicis, and the spinous processes of C3–T6.

- Usual innervation: Data not available.

Semispinalis cervicis
- Usual attachments: From the articular processes of C3–C7, transverse processes of T1–T7, and spinalis to the spine of the axis and spinous processes of C3–C7.
- Usual innervation: Data not available.
- Notes: Raven (1950) did not describe a distinct **semispinalis thoracis** in the gorilla specimen dissected by him; we found this latter muscle in our VU GG1 specimen (see Fig. 52).

Semispinalis capitis (Fig. 51)
- Usual attachments: From the articular processes of C2–C7 and the transverse processes of T1–T5 to the occipital bone.
- Usual innervation: Dorsal rami of C2 and C3, but also of C4–C8 plus T1–T3 (Raven 1950, Gibbs 1999).

Multifidus
- Usual attachments: From the sacroiliac region, mammillary processes of the lumbar vertebrae, the transverse processes of the thoracic vertebrae, and the articular processes of C4–C7, to the spinous processes of the lumbar, thoracic and cervical vertebrae except the atlas.
- Usual innervation: Data not available.

Rotatores (Fig. 52)
- Usual attachments: From the transverse processes to the spinous processes of the vertebrae, the rotatores breves going from one vertebra to the next, and the rotatores longi bridging over one vertebra.
- Usual innervation: Data not available.
- Synonymy: Rotatores breves and longi (Gibbs 1999).

Interspinalis
- Usual attachments: Connects the spinous processes of the adjacent vertebrae, from C3 to T2.
- Usual innervation: Data not available.

Intertransversarii
- Usual attachments: Similar to those of modern humans, according to Raven (1950) and Gibbs (1999).
- Usual innervation: Data not available.

CHAPTER 6

Diaphragmatic and Abdominal Musculature

Diaphragma (Fig. 56)
- Usual attachments: From the posterior surface of the xiphoid process, ribs 7–13, and the tendinous raphe between the last rib and L1, to the centrum tendineum.
- Usual innervation: Phrenic nerves, from C3, C4 and C5 (Raven 1950, Gibbs 1999).

Rectus abdominis (Figs. 53, 54, 55)
- Usual attachments: From the xiphisternum and costal cartilages 5–9 to the ventral spine of pubis and symphysial ligament.
- Usual innervation: Intercostal nerves 1–13, and possibly also by the first lumbar nerve (Raven 1950, Gibbs 1999).
- Notes: According to Raven (1950), the **pyramidalis** was not present as a distinct muscle in the gorilla dissected by him.

Tensor linea semilunaris
- Usual attachments: From pubis to aponeurosis of the transversus abdominis and of obliquus abdominis internus.
- Usual innervation: Data not available.
- Notes: As noted by Gibbs (1999) this muscle has not been described in modern humans, hylobatids, orangutans and chimpanzees.

Cremaster
- Usual attachments: Originates from obliquus internus abdominis and contains a contribution from the transversus abdominis.
- Usual innervation: Data not available.

Obliquus externus abdominis (Figs. 15, 53)
- Usual attachments: From ribs 5–13 to sheath of rectus abdominis, the inguinal region, and the anterior superior iliac spine.
- Usual innervation: Data not available.

Obliquus internus abdominis
* Usual attachments: From the iliac crest, thoracolumbar fascia and anterior superior iliac spine, to the last four ribs, pubis and aponeurosis of the obliquus externus abdominis.
* Usual innervation: At least by some branches of the intercostal nerves 1–13, and possibly also by the first lumbar nerve (Raven 1950, Gibbs 1999).

Transversus abdominis (Figs. 54, 55)
* Usual attachments: From the last six costal cartilages, thoracolumbar fascia, iliac crest and anterior superior iliac spine to linea alba.
* Usual innervation: Intercostal nerves 8–13 and first lumbar nerve (Raven 1950, Gibbs 1999).

Quadratus lumborum
* Usual attachments: From the iliac crest and iliolumbar ligament to T13, L1–2 and rib 13.
* Usual innervation: Ventral rami of T13–L2 (Raven 1950, Gibbs 1999).

CHAPTER 7

Perineal, Coccygeal and Anal Musculature

Coccygeus
- Usual attachments: From the ischium and ilium to the coccyx, the inferior part of the sacrum and the anococcygeal raphe.
- Usual innervation: S3 (Raven 1950, Gibbs 1999).

Flexor caudae
- Usual attachments: From the last sacral vertebra to the middorsal fascia, being mainly a 'vestigial' muscle.
- Usual innervation: S5 (Raven 1950).

Iliococcygeus
- Usual attachments: From the obturator fascia to the coccyx and in some specimens also to the sacrum according to the literature review of Gibbs (1999).
- Usual innervation: S1, S2 and S3 (Raven 1950, Gibbs 1999).

Levator ani
- Usual attachments: From the obturator fascia to the anococcygeal raphe and coccyx.
- Usual innervation: S1, S2 and S3 (Raven 1950, Gibbs 1999).

Pubovesicalis
- Usual attachments: From the pubis to the bladder.
- Usual innervation: Data not available.
- Synonymy: Ligamentum puboprostaticum or puboampullaris (Gibbs 1999).

Pubococcygeus
- Usual attachments: From the pubis and obturator fascia to the wall of the rectum and coccyx.
- Usual innervation: S1, S2 and S3 (Raven 1950, Gibbs 1999).

Puborectalis
- Usual attachments: From the connective tissue in the region of the symphyseal angle to deep external anal sphincter, forming a puborectal sling according to the literature review of Gibbs (1999).
- Usual innervation: Data not available.

Sphincter ani externus
- Usual attachments: From the ischium to the anococcygeal raphe and to bulbospongious.
- Usual innervation: S1, S2 and S3 (Raven 1950, Gibbs 1999).

Bulbospongiosus
- Usual attachments: From the perineal body, ischium and penile bulb to the crura, to a fibrous expansion on the penis, and to inferior surface of penis.
- Usual innervation: Data not available.
- Synonymy: Bulbocavernosus (Raven 1950).

Ischiocavernosus
- Usual attachments: From the ischium to the corpora cavernosum.
- Usual innervation: Data not available.

Sphincter urethrae
- Usual attachments: According to the literature review of Gibbs (1999), in gorillas this is a true sphincter, with no invasion of the prostate, similar to that found in prepubescent modern humans *Homo*.
- Usual innervation: Data not available.

Transversus perinei profundus
- Usual attachments: According to the literature review of Gibbs (1999), in gorillas this muscle originates from the fascia just above the ischial ramus, being continuous with the sphincter urethrae, and interdigitating at the midline with the external anal sphincter, the bulbospongious and the smooth muscle of the rectal wall; according to Gibbs (1999), this structure may be analogous to a perineal body.
- Usual innervation: Data not available.

Transversus perinei superficialis
- Usual attachments: From the ischium to the bulbospongiosus and sphincter ani externus, being absent in at least some gorillas according to the literature review of Gibbs (1999).
- Usual innervation: Data not available.

Rectococcygeus
- Usual attachments: From the last coccygeal vertebra and fascia of middorsal line posterior to this structure, to the rectum, being present in half of the gorillas according to the literature review of Gibbs (1999).
- Usual innervation: Nerve to pubococcygeus (Raven 1950, Gibbs 1999).
- Synonymy: Caudoanalis (Raven 1950).

Rectourethralis

- Usual attachments: According to the literature review of Gibbs (1999) this is a smooth muscle running from the inferior part of each side of the rectal wall to the sphincter urethrae, dorsal urethra and corpus spongiosum.
- Usual innervation: Nerve to pubococcygeus (Raven 1950, Gibbs 1999).
- Notes: To our knowledge, the muscles **rectovesicalis, regionis analis, regionis urogenitalis, compressor urethrae, sphincter urethrovaginalis, sphincter pyloricus, suspensori duodeni, sphincter ani internus, sphincter ductus choledochi, sphincter ampullae, detrusor vesicae, trigoni vesicae, vesicoprostaticus, vesicovaginalis, puboprostaticus** and **rectouterinus** were not described in detail in gorillas, and we could also not investigate these muscles in detail in the gorillas dissected by us, mainly because in some of these specimens the regions where these muscles are were removed/damaged before our dissections.

CHAPTER 8

Pelvic and Lower Limb Musculature

Iliacus (Figs. 59, 60)
- Usual attachments: From the iliac fossa and lumbodorsal fascia to the lesser trochanter of the femur.
- Usual innervation: Branches of the femoral nerve (Raven 1950, Gibbs 1999).

Psoas major (Figs. 59, 60)
- Usual attachments: From the ilium and T12, as well as from all lumbar vertebrae and, at least sometimes, also from S1 according to the literature review of Gibbs (1999); to the lesser trochanter of the femur.
- Usual innervation: Branches of the femoral nerve (Raven 1950), and also of the first two or three lumbar nerves according to the literature review of Gibbs (1999).

Psoas minor
- Usual attachments: From T12–T13, and at least sometimes from T11, to the iliopubic eminence and pectineal line.
- Usual innervation: Branches of the femoral nerve (Raven 1950); L1 and, occasionally, also by T12 (Gibbs 1999).

Gluteus maximus (Figs. 57, 58)
- Usual attachments: From the thoracolumbar fascia, sacrum, iscium, posterior iliac crest, coccyx, sacrotuberal ligament and/or posterior superior iliac spine, to the lateral aspect of the femur.
- Usual innervation: Inferior gluteal nerve (Raven 1950, Gibbs 1999).

Gluteus medius (Figs. 57, 58)
- Usual attachments: From the fascia lata, ilium and sacrum, to the greater trochanter of the femur.
- Usual innervation: Superior gluteal nerve (Raven 1950, Gibbs 1999).

Gluteus minimus (Fig. 58)
- Usual attachments: From the ilium and fascia lata to the greater trochanter and neck of the femur.
- Usual innervation: Superior gluteal nerve (Raven 1950, Gibbs 1999).

Scansorius
- Usual attachments: From the fascia lata to the lateral surface of the femur.
- Usual innervation: Superior gluteal nerve (Raven 1950).
- Notes: In some gorillas the scansorius is not present as a distinct muscle (see Gibbs 1999).

Gemellus inferior (Fig. 58)
- Usual attachments: From the body of the ischium to the trochanteric fossa of the femur.
- Usual innervation: Branch of the sacral plexus from the base of the sciatic nerve (Raven 1950, Gibbs 1999).
- Notes: The **gemellus superior** is absent in most of the gorillas (see Gibbs 1999).

Obturatorius internus (Fig. 58)
- Usual attachments: From the ilium, margin of obturator foramen, obturator membrane and ischium to the trochanteric fossa of the femur.
- Usual innervation: S1, S2, S3, and possibly L3 (Raven 1950, Gibbs 1999).
- Synonymy: Obturator internus (Raven 1950, Gibbs 1999).

Obturatorius externus
- Usual attachments: From the ilium, pubis, obturator membrane and ischium to the trochanteric fossa of the femur and, occasionally, to the superior femoral neck according to the literature review of Gibbs (1999).
- Usual innervation: Obturator nerve (Raven 1950, Gibbs 1999).
- Synonymy: Obturator externus (Raven 1950, Gibbs 1999).

Piriformis (Fig. 58)
- Usual attachments: From the sacrum and, occasionally, from the coccyx according to Raven (1950); to the greater trochanter of the femur.
- Usual innervation: S1, S2 (Raven 1950, Gibbs 1999).

Quadratus femoris (Fig. 58)
- Usual attachments: From the corpus ischium to the greater and lesser trochanters of the femur and to the region between these two latter structures.
- Usual innervation: Branch of the sacral plexus, from the base of the sciatic nerve, with the nerve to gemellus inferior (Raven 1950, Gibbs 1999).

Articularis genus
- Usual attachments: From the distal portion of the femur to the lateroproximal portion of the articular capsule of the knee joint.
- Usual innervation: Branch of the femoral nerve (Raven 1950, Gibbs 1999).

Rectus femoris (143.0 g; Figs. 59, 60)
- Usual attachments: From the anterior inferior iliac spine (caput rectum) and ilium superior to the acetabulum (caput reflexum) to patella.
- Usual innervation: Femoral nerve (Raven 1950, Gibbs 1999).
- Notes: According to the literature review of Gibbs (1999), the caput rectum and caput reflexum are present as distinct structures in 2/3 of gorillas.

Vastus intermedius (vastus intermedius + lateralis + medialis: 552.0 g)
- Usual attachments: From the ventral and medial aspects of the femoral shaft and, at least sometimes, from the femoral neck (see, e.g., Raven 1950), to the patella.
- Usual innervation: Femoral nerve (Raven 1950, Gibbs 1999).

Vastus lateralis (vastus intermedius + lateralis + medialis: 552.0 g; Figs. 59, 60)
- Usual attachments: From the greater trochanter, the femoral shaft just distal to this latter structure, and the lateral aspect of the femoral shaft, to the patella, the tendon of rectus femoris, and the capsule of the knee joint.
- Usual innervation: Femoral nerve (Raven 1950, Gibbs 1999).

Vastus medialis (vastus intermedius, lateralis and medialis: 552.0 g; Figs. 59, 60)
- Usual attachments: From the medial and dorsal aspects of the femur to the patella and the medial patellar ligament.
- Usual innervation: Femoral nerve (Raven 1950, Gibbs 1999).

Sartorius (98.5 g; Fig. 59)
- Usual attachments: From the lateral border of the ilium to the medial side of the cnemial crest of the tibia, and occasionally to the crural fascia medial to the tibial tuberosity according to the literature review of Gibbs (1999).
- Usual innervation: Femoral nerve (Raven 1950, Gibbs 1999).

Tensor fasciae latae
- Usual attachments: From the anterior superior iliac spine and fascia lata just distal to it, to the iliotibial tract, specifically to the fascia lata lateral to the vastus lateralis according to Raven (1950).
- Usual innervation: Superior gluteal nerve (Raven 1950, Gibbs 1999).

Adductor brevis (adductor brevis, longus, magnus and minimus: 1125.5 g; Fig. 60)
- Usual attachments: From the body, inferior ramus and, occasionally, from the superior ramus (see Gibbs 1999) of the pubis to the dorsal aspect of the femur.
- Usual innervation: Ventral division of the obturator nerve (Raven 1950, Gibbs 1999).
- Notes: A division of this muscle into two parts is present in 1/3 of gorillas according to the literature review of Gibbs (1999). Such a division was present in our VU GG1 specimen (see Fig. 60).

Adductor longus (adductor brevis, longus, magnus and minimus: 1125.5 g; Fig. 59)
- Usual attachments: From the superior ramus of the pubis to the dorsolateral aspect of the femur.
- Usual innervation: Ventral division of the obturator nerve (Raven 1950, Gibbs 1999).

Adductor magnus (adductor brevis, longus, magnus and minimus: 1125.5 g; Fig. 59)
- Usual attachments: From tuberosity of the ischium to the medial epicondyle of the femur (caput longum), and from inferioris ramus of the pubis and ischium to the common tendon of adductor longus and adductor brevis and the dorsal aspect of the femur (caput breve).
- Usual innervation: Caput longum by the tibial nerve or by the flexores femoris nerve, and caput breve by the obturator nerve (Raven 1950, Gibbs 1999). According to Gibbs (1999), the tibial nerve innervates the long head in 1/2 of the gorillas.

Adductor minimus (adductor brevis, longus, magnus and minimus: 1125.5 g)
- Usual attachments: From the superior and inferior pubic rami and ischium to the dorsal surface of the femur just distal and lateral to the lesser trochanter of this bone.
- Usual innervation: Obturator nerve (Raven 1950).

Gracilis (279.0 g; Fig. 59)
- Usual attachments: From the inferior and inferior pubic rami and the pubic body to the medial portion of the tibia.
- Usual innervation: Obturator nerve (Raven 1950), and specifically by a ventral branch of this nerve according to the literature review of Gibbs (1999).

Pectineus (48.0 g; Fig. 59)
- Usual attachments: From the superior ramus of the pubis to the dorsal surface of the femur directly distal to the lesser trochanter of this latter bone.
- Usual innervation: Femoral nerve; not by both the obturator and femoral nerves as is occasionally the case in *Pan* and *Pongo* (Raven 1950, Gibbs 1999).

Iliocapsularis
- Usual attachments: This muscle was found in the adult gorilla described by Raven (1950), and also in an infant gorilla dissected by him, running from the lateral border of the ilium to the base of the lesser trochanter of the femur and a small area adjacent to it.
- Usual innervation: Data not available.

Biceps femoris
- Usual attachments: From ischial tuberosity to the ventroproximal portion of the tibia, fibula and distal portion of the iliotibial tract (caput longum); from

the dorsolateral aspect of the femur to the head of the fibula and the deep fascia of the leg (caput breve).

- Usual innervation: Caput longum is innervated by the flexores femoris nerve, as is the case in other apes, or by the tibial nerve, as is the case in modern humans: innervation by the tibial nerve was described by Raven (1950), and occurs in 1/2 of the gorillas according to Gibbs (1999); caput breve is innervated by the common peroneal nerve, as is the case in other apes and in modern humans (Raven 1950, Gibbs 1999).

Semimembranosus (206.0 g)
- Usual attachments: From ischial tuberosity to the infracondylar ridge on the medial side of the tibial head, and also to the popliteal fascia and the knee capsule via the oblique popliteal ligament according to the literature review of Gibbs (1999).
- Usual innervation: By the flexores femoris nerve, as is the case in other apes, or by the tibial nerve, as is the case in modern humans: innervation by the tibial nerve was described by Raven (1950), and occurs in 1/2 of the gorillas according to the literature review of Gibbs (1999).

Semitendinosus (333.2 g)
- Usual attachments: From ischial tuberosity to the ventral surface of the tibia and the fascia of the leg.
- Usual innervation: By the flexores femoris nerve, as is the case in other apes, or by the tibial nerve, as is the case in modern humans: innervation by the tibial nerve was described by Raven (1950), and occurs in 1/2 of the gorillas according to Gibbs (1999).

Extensor digitorum longus (61.6 g; Fig. 61)
- Usual attachments: From the lateral condyle of the tibia, capitulum fibulae, crista anterior fibulae and the interosseous membrane to the distal phalanges of digits 2–4.
- Usual innervation: Deep peroneal nerve (Raven 1950, Gibbs 1999).

Extensor hallucis longus (25.8 g; Fig. 61)
- Usual attachments: From the interosseous membrane and the medial surface of the fibula to the proximal, and mainly to the distal phalanges of digit 1.
- Usual innervation: Deep peroneal nerve (Raven 1950, Gibbs 1999).
- Notes: According to the literature review of Gibbs (1999), the **fibularis tertius** ('peroneus tertius') is present in 30% of the gorillas (when it is present in the primates, it usually has a fascial insertion onto metacarpal V), 5% of Pan, 1/2 of *Hylobates*, and 95% of modern humans, but is absent in orangutans.

Tibialis anterior (127.0 g; Figs. 61, 62)
- Usual attachments: From the medial side of the lateral condyle and the medial side of the shaft of the tibia and from the interosseous membrane to the plantar aspect of the base of metatarsal I (smaller tendon) and to the plantar side of

the lateral cuneiform (larger tendon). These two tendons were clearly present in our VU GG1 specimen.

- Usual innervation: Deep peroneal nerve (Raven 1950, Gibbs 1999).

Fibularis brevis (40.3 g; Fig. 61)
- Usual attachments: From the lateral margin of the fibula to metacarpal V and also to the ligamentous expansion of the tendon of extensor digitorum at the metatarsophalangeal joint of digit 5, forming a 'fibularis or peroneus digiti quinti' *sensu* Raven (1950); a small tendon to the proximal and middle phalanges of digit 5 is occasionally present in gorillas, as well as in modern humans (e.g., Macalister 1873, Raven 1950, Gibbs 1999).
- Usual innervation: Superficial peroneal nerve (Raven 1950, Gibbs 1999).
- Synonymy: Peroneous brevis or peronaeus brevis (Raven 1950, Gibbs 1999).

Fibularis longus (69.6 g; Fig. 61)
- Usual attachments: From the head and lateral margin of the fibula to the malleolus and mainly to the base of metatarsal I.
- Usual innervation: Superficial peroneal nerve (Raven 1950, Gibbs 1999).
- Synonymy: Peroneous longus or peronaeus longus (Raven 1950, Gibbs 1999).

Gastrocnemius (190.9 g; Fig. 62)
- Usual attachments: From the medial (caput mediale) and lateral (caput laterale) condyles of the femur, and often also from the knee capsule, to calcaneal tuberosity.
- Usual innervation: Tibial nerve (Raven 1950, Gibbs 1999).
- Notes: According to the literature review of Gibbs (1999), the **plantaris** is absent in gorillas and hylobatids, while it is present in a few orangutans and in about 3/5 of *Pan* and in 90–95% of humans.

Soleus (159.5 g; Fig. 63)
- Usual attachments: From the head and proximal margin of the shaft of the fibula to the calcaneal tuberosity; according to Chapman (1878), Duvernoy (1855-1856), Raven (1950) and Gibbs (1999), the origin from the tibia is absent, or 'rudimentary', in gorillas.
- Usual innervation: Tibial nerve (Raven 1950, Gibbs 1999).

Flexor digitorum longus (86.4 g; Figs. 64, 68)
- Usual attachments: From the dorsal aspect of the tibia to the distal phalanges of digits 2–5; according to Gibbs (1999), an insertion on digit 5, digit 4, digit 3 and digit 2 is present in 6/6, 3/6, 3/6 and 6/6 of the gorillas described in the literature, respectively. In our VU GG1 specimen there was also effectively a tendon to digit 5.
- Usual innervation: Tibial nerve (Raven 1950, Gibbs 1999).

Flexor hallucis longus (177.0 g; Figs. 64, 65, 66, 67, 68)
• Usual attachments: From the dorsal surface of the fibula and the interosseous membrane to the distal phalanges of digit 1 (this part corresponds to the flexor hallucis longus *sensu* stricto) and of digits 3 and 4 and, occasionally, also 2 and/or 5 (this part becomes fused with the tendons of the flexor digitorum longus); according to the literature review of Gibbs 1999, an insertion to digits 3 and 4 is often present in gorillas; an insertion to digit 5 was reported by, e.g., Chapman (1878) and an insertion to digit 2 was described by Raven (1950). In our VU GG1 specimen the flexor hallucis longus inserted onto digits 1, 2, 3, 4 and 5.
• Usual innervation: Tibial nerve (Raven 1950, Gibbs 1999).

Popliteus (Fig. 64)
• Usual attachments: From the lateral epicondyle of the femur and the capsule of the knee joint to the dorsal surface of the proximal tibia.
• Usual innervation: Tibial nerve (Raven 1950, Gibbs 1999).

Tibialis posterior (58.2 g)
• Usual attachments: From the lateral aspect of the tibia, the medial aspect of the fibula, and the interosseous membrane to malleolus, navicular tuberosity, sheath of fibularis longus, and lateral cuneiform.
• Usual innervation: Tibial nerve (Raven 1950, Gibbs 1999).

Extensor digitorum brevis (14.2 g; Fig. 61)
• Usual attachments: From the lateral surface of calcaneus to the middle and distal phalanges of digits 2, 3 and 4, together with the tendons of the extensor digitorum longus.
• Usual innervation: Deep peroneal nerve (Raven 1950, Gibbs 1999).

Extensor hallucis brevis (5.1 g; Fig. 61)
• Usual attachments: From the calcaneus and cruciate ligament to the proximal phalanx of digit 1. In the specimen dissected by Raven (1950) there was an accessory slip running from the lateral part of the main body of the muscle to the proximal phalanx of digit 2 (such a slip is also occasionally present in humans: see Raven 1950, Gibbs 1999).
• Usual innervation: Deep peroneal nerve (Raven 1950, Gibbs 1999).

Abductor digiti minimi (Figs. 65, 66, 68)
• Usual attachments: From the medial and lateral calcaneus to metatarsal V and proximal phalanx of digit 5.
• Usual innervation: Lateral plantar nerve (Raven 1950, Gibbs 1999).
• Synonymy: Abductor digiti quinti (Raven 1950).

Abductor hallucis (34.4 g; Figs. 65, 66, 67, 68)
- Usual attachments: From the calcaneus and cuboid to metatarsal I and the base of the proximal phalanx of digit 1; the slip to the base of metatarsal I is often called '**abductor ossis metacarpi hallucis**' in gorillas.
- Usual innervation: Medial plantar nerve (Raven 1950, Gibbs 1999).

Abductor metatarsi quinti (weight not measured)
- Usual attachments: From the lateral border of the calcaneus to the peroneal tubercle of metatarsal V.
- Usual innervation: Lateral plantar nerve (Raven 1950, Gibbs 1999).
- Notes: This muscle has been described in great apes as quite distinct from the abductor digiti minimi, and has been reported as a variant in modern humans (see Gibbs 1999).
- Synonymy: Abductor ossis metacarpi digiti quinti (Raven 1950); abductor os metatarsi digiti minimi (Gibbs 1999).

Flexor digitorum brevis (28.0 g; Figs. 65, 66, 67)
- Usual attachments: Caput superficiale from the calcaneus to the middle phalanges of digits 2 and 3 and to the tendon of caput profundum. Caput profundum from plantar surface of tendon of flexor digitorum longus to digits 4 and 5 according to Raven (1950), although Gibbs (1999) suggested that in gorillas and other apes digit 4 is usually supplied by the caput superficiale, and not by the caput profundum. In our VU GG1 specimen the caput superficiale appeared to go to digits 2 and 3, while the caput profundum apparently attached to digit 4. An occasional origin of the caput profundum from the leg has been reported by Straus (1930).
- Usual innervation: Medial plantar nerve (Raven 1950, Gibbs 1999).

Quadratus plantae (weight not measured; Fig. 67)
- Usual attachments: From the calcaneus to the lateral border of the tendon of the flexor digitorum longus; according to Gibbs (1999), this represents the caput laterale of modern humans, the caput mediale is usually present in orangutans but absent in African apes.
- Usual innervation: Lateral plantar nerve (Raven 1950, Gibbs 1999).
- Synonymy: Flexor accessorius (Gibbs 1999).

Lumbricales (I: 2.44 g; II: 2.45 g; III: 2.10 g; 4: 0.91 g; Fig. 68)
- Usual attachments: Mainly from common tendons of flexor digitorum longus and flexor hallucis longus and the proximal part of their individual tendons, to the medial side of proximal phalanges and extensor aponeuroses of digits 2 (lumbricalis I), 3 (lumbricalis II), 4 (lumbricalis III) and 5 (lumbricalis IV).
- Usual innervation: Medial plantar nerve (lumbricalis I) and lateral plantar nerve (lumbricales II, III and IV) (Raven 1950, Gibbs 1999).

Adductor hallucis (caput transversum: 23.3 g; caput obliquum: 20.9 g; Figs. 61, 65, 66, 67, 68)
- Usual attachments: Caput transversum originates from the aponeurotic sheet connected to metatarsals II, III and IV and to the deep dorsal fascia of the foot, and inserts onto the capitulum of metatarsal I, metatarsophalangeal joint, shaft of the proximal phalanx and the interphalangeal joint of digit 1. Caput obliquum originates from bases of metatarsals II and III and adjacent ligaments and inserts onto the capitulum and capsule of metatarsal I just proximal to the insertion of the caput transversum.
- Usual innervation: Deep branch of lateral plantar nerve (Raven 1950, Gibbs 1999).
- Notes: In modern humans the '**transversalis pedis**' (not listed in Terminologia Anatomica 1998) corresponds to the caput transversum of the adductor hallucis.

Flexor digiti mimini brevis (Figs. 65, 66, 67, 68)
- Usual attachments: From the bases of metatarsals IV and V and their interosseous ligaments to the base of the proximal phalanx of digit 5.
- Usual innervation: Lateral plantar nerve (Raven 1950, Gibbs 1999).
- Synonymy: Flexor digiti quinti brevis (Raven 1950); flexor digiti minimi (Gibbs 1999).

Flexor hallucis brevis (15.9 g; Figs. 65, 66, 67, 68)
- Usual attachments: The caput mediale from the calcaneus, the second cuneiform, first tarsometatarsal joint and, sometimes, also from the navicular bone (see Gibbs 1999), to the head of metatarsal I and the metatarsophalangeal joint and the base of the proximal phalanx of digit 1. The caput laterale from metatarsal I and adjacent ligaments to the metatarsophalangeal joint of digit 1.
- Usual innervation: Medial plantar nerve (Raven 1950, Gibbs 1999).

Opponens digiti minimi
- Usual attachments: From the base of metatarsal V and an adjacent sesamoid bone to the lateral border of the shaft of metatarsal V.
- Usual innervation: Lateral plantar nerve (Raven 1950).
- Notes: This structure is usually considered to be a deep fascicle of the flexor digiti minimi in orangutans and modern humans (it was not listed in Terminologia Anatomica 1998), but has been described as a distinct muscle in *Pan* and *Gorilla* (see Raven 1950, Gibbs 1999).
- Synonymy: Opponens digiti quinti (Raven 1950).

Interossei dorsales (I: 8.0 g; II: 5.0 g; III: 2.9 g; IV: 4.2 g; Fig. 61)
- Usual attachments: Interosseous dorsalis I mainly from the medial cuneiform and metatarsal II to the medial aspect of the base of the proximal phalanx of digit 2; interosseous dorsalis II mainly from metatarsals II and III to the medial

aspect of the base of the proximal phalanx of digit 3; interosseous dorsalis III mainly from metatarsals III and IV to the lateral aspect of the base of the proximal phalanx of digit 3; interosseous dorsalis IV mainly from metatarsals IV and V and the adjacent sesamoid bone to the lateral aspect of the base of proximal phalanx of digit 4.

- Usual innervation: Deep branch of lateral plantar nerve (Raven 1950, Gibbs 1999).

Interossei plantares (I: 4.7 g; II: 5.5 g; III: weight not measured)

- Usual attachments: Interosseous plantaris I mainly from metatarsals II and III to the lateral side of the proximal phalanx and of the metatarsophalangeal joint of digit 2; interosseous plantaris II mainly from metatarsal IV and the adjacent sesamoid bone to the medial side of the proximal phalanx and of the metatarsophalangeal joint of digit 4; interosseous plantaris III mainly from metatarsal V and the adjacent sesamoid bone to the medial side of the proximal phalanx and of metatarsophalangeal joint of digit 5.
- Usual innervation: Deep branch of the lateral plantar nerve (Raven 1950, Gibbs 1999).

Appendix I
Literature Including Information about the Muscles of Gorilla*

Aiello L, Dean C. 1990. An introduction to human evolutionary anatomy. San Diego: Academic Press.

Ashton EH, Oxnard CE. 1963. The musculature of the primate shoulder. Trans Zool Soc Lond 29: 553–650.

Ashton EH, Oxnard CE. 1964. Functional adaptations of the primate shoulder girdle. Proc Zool Soc Lond 142: 49–66.

Atkinson WB, Elftman H. 1950. Female reproductive system of the gorilla. In The Anatomy of the Gorilla (Gregory WK,Ed.). New York: Columbia University Press. pp. 205–211.

Aziz MA, Dunlap SS. 1986. The human extensor digitorum profundus muscle with comments on the evolution of the primate hand. Primates 27: 293–319.

Barnard WS. 1875. Observations on the membral musculation of *Simia satyrus* (Orang) and the comparative myology of man and the apes. Proc Amer Assoc Adv Sci 24: 112–144.

Beddard FE. 1893. Contributions to the anatomy of the anthropoid apes. Trans Zool Soc Lond 13: 177–218.

Bischoff TLW. 1880. Beitrage zur Anatomie des *Gorilla*. Abh Bayer Akad Wiss Miinchen Math Phys Kl 13: 1–48.

Bojsen–Møller F. 1978. Extensor carpi radialis longus muscle and the evolution of the first intermetacarpal ligament. Am J Phys Anthropol 48: 177–184.

Broca P. 1869. L'ordre des primates—parallele anatomique de l'homme et des singes. Bull Soc Anthropol Paris 4: 228–401.

Brooks HSJ. 1886a. On the morphology of the intrinsic muscles of the little finger, with some observations on the ulnar head of the short flexor of the thumb. J Anat Physiol 20: 644–661.

Brooks HSJ. 1886b. Variations in the nerve supply of the flexor brevis pollicis muscle. J Anat Physiol 20: 641–644.

Brooks HSJ. 1887. On the short muscles of the pollex and hallux of the anthropoid apes, with special reference to the opponens hallucis. J Anat Physiol 22: 78–95.

Cave AJE. 1979. The mammalian temporo-pterygoid ligament. J Zool Lond 188: 517–532.

Chapman HC. 1878. On the Structure of the *Gorilla*. Proc Acad Nat Sci Philad 30: 385–394.

Chudzinski T. 1885. Sur les muscles peaussiers du crane et de la face observes sur un jeune gorilla. Bull Soc Anthropol Paris 8: 583–586.

*List not exhaustive

Chylewski W. 1926. Glowa lokciowa m. nawrotnego oblego (m. pronator teres) w szerega naczelynych. C R Soc Sci Varsowie, Cl 3, 19: 367–375.

Day MH, Napier J. 1963. The functional significance of the deep head of flexor pollicis brevis in primates. Folia Primatol 1: 122–134.

Dean MC. 1984. Comparative myology of the hominoid cranial base, I, the muscular relationships and bony attachments of the digastric muscle. Folia Primatol 43: 234–48.

Dean MC. 1985. Comparative myology of the hominoid cranial base, II, the muscles of the prevertebral and upper pharyngeal region. Folia Primatol 44: 40–51.

Deniker J. 1885. Recherches anatomiques et embryologiques sur les singes anthropoïdes—foetus de gorille et de gibbon. Arch Zool Exp Gén 3, suppl 3: 1–265.

Delrich TM. 1978. Pelvic and perineal anatomy of the male gorilla: selected observations. Anat Record 191: 433–446.

Diogo R, Abdala V, Lonergan N, Wood BA. 2008. From fish to modern humans—comparative anatomy, homologies and evolution of the head and neck musculature. J Anat 213: 391–424.

Diogo R, Abdala V, Aziz MA, Lonergan N, Wood BA. 2009a. From fish to modern humans—comparative anatomy, homologies and evolution of the pectoral and forelimb musculature. J Anat 214: 694–716.

Diogo R, Wood BA, Aziz MA, Burrows A. 2009b. On the origin, homologies and evolution of primate facial muscles, with a particular focus on hominoids and a suggested unifying nomenclature for the facial muscles of the Mammalia. J Anat 215: 300–319.

Diogo R, Abdala V. 2010. Muscles of vertebrates—comparative anatomy, evolution, homologies and development.Enfield: Science Publishers.

Duckworth WLH. 1898. Note on a foetus of *Gorilla* savagei. J Anat Physiol 33: 82–90.

Duckworth WLH. 1904. Studies from the Anthropological Laboratory, the Anatomy School, Cambridge.London: C. J. Clay & Sons.

Duckworth WLH. 1912. On some points in the anatomy of the plica cocalis. J Anat Physiol 47: 80–115.

Duckworth WLH. 1915. Morphology and anthropology (2nd ed.). Cambridge: Cambridge University Press.

Duvernoy M. 1855–1856. Des caracteres anatomiques de grands singes pseudoanthropomorphes anthropomorphes. Arch Mus Natl Hist Nat Paris 8: 1–248.

Dylevsky I. 1967. Contribution to the ontogenesis of the flexor digitorum superficialis and the flexor digitorum profundus in man. Folia Morphol (Praha) 15: 330–335.

Edgeworth FH. 1935. The cranial muscles of vertebrates.Cambridge: University Press.

Elftman HO. 1932. The evolution of the pelvic floor of primates. Am J Anat 51: 307–346.

Elftman H, Atkinson WB. 1950. The abdominal viscera of the gorilla. In The Anatomy of the Gorilla (Gregory WK Ed.). New York: Columbia University Press. pp. 197–201.

Ehlers E. 1881. Beiträge zur Kenntnis des gorilla und chimpanse. Abh d K Gesell d Wissensch du Göttingen, Phys Cl 38: 3–77.

Eisler P. 1890. Das Gefäss—und periphere Nervensystem des *Gorilla*. Talle: Tausch and Grosse.

Frey H. 1913. Der Musculus triceps surae in der Primatenreihe. Morph Jahrb 47: 1–192.

Gerhardt U. 1906. Die Morphologie des Urogenitalsystems eines weiblichen *Gorilla*. Z Gesammten Naturewissenschaften 41: 632–654.

Giacomini C. 1897. 'Plica semilunaris' et larynx chez les singes anthropomorphes. Arch Ital Biol 28: 98–119.

Gibbs S. 1999. Comparative soft tissue morphology of the extant Hominoidea, including Man. Unpublished PhD Thesis, The University of Liverpool, Liverpool.

Gibbs S, Collard M, Wood BA. 2000. Soft-tissue characters in higher primate phylogenetics. Proc Natl Acad Sci US 97: 11130–11132.

Gibbs S, Collard M, Wood BA. 2002. Soft-tissue anatomy of the extant hominoids: a review and phylogenetic analysis. J Anat 200: 3–49.

Göllner K. 1982. Untersuchungen über die vom N. trigeminus innervierte Kiefermusculatur des Schimpansen (*Pan troglodytes*, Blumenbach 1799) und des Gorilla (*Gorilla gorilla gorilla*, Savage and Wyman 1847). Gegen Morphol Jahrb 128: 851–903.

Groves CP. 1986. Systematics of the great apes. In Comparative Primate Biology: Systematics, Evolution and Anatomy, Vol. 1 (Swindler DR, Erwin J Eds.). New York: A.R. Liss. pp. 187–217.

Groves CP. 1995. Revised character descriptions for Hominoidea. Typescript .

Hamada Y. 1985. Primate hip and thigh muscles: comparative anatomy and dry weights. In Primate Morphophysiology, Locomotor Analyses and Human Bipedalism (Kondo S Ed.). Tokyo: University of Tokyo Press. pp. 131–152.

Hartmann R. 1886. Anthropoid apes. London: Keegan.

Hepburn D. 1892. The comparative anatomy of the muscles and nerves of the superior and inferior extremities of the anthropoid apes, I, Myology of the superior extremity. J Anat Physiol 26: 149–186.

Hepburn D. 1896. A revised description of the dorsal interosseous muscles of the human hand, with suggestions for a new nomenclature of the palmar interosseous muscles and some observations on the corresponding muscles in the anthropoid apes. Trans R Soc Edin 38: 557–565.

Hill WCO. 1949. Some points in the enteric anatomy of the great apes. Proc Zool Soc London 119: 19–32.

Hill WCO, Harrison-Matthews L. 1949. The male external genitalia of the gorilla, with remarks on the os penis of other Hominoidea. Proc Zool Soc London 119: 363–378.

Hill WCO, Harrison-Matthews L. 1950. Supplementary note on the male external genitalia of *Gorilla*. Proc Zool Soc London 120: 311–316.

Hilloowala RA. 1980. The migrating omohyoid muscle—its significance? J Human Evol 9: 165–172.

Hofër 1892. Vergleichend-anatomische Studien uber die Nerven des Armes und der Hand bei den Affen und dem Menschen. Munchener Med Abhandl 30: 1–106.

Hosokawa H, Kamiya T. 1961–1962. Anatomical sketches of visceral organs of the mountain gorilla (*Gorilla gorilla beringei*). Primates 3: 1–28.

Huber E. 1930. Evolution of facial musculature and cutaneous field of trigeminus—Part II. Q Rev Biol 5: 389–437.

Huber E. 1931. Evolution of facial musculature and expression. Baltimore: The Johns Hopkins University Press.

Huxley TH. 1864. The structure and classification of the Mammalia. Med Times Gazette 1864: 398–468.

Huxley TH. 1871. The anatomy of vertebrated animals. London: J. & A. Churchill.

Jouffroy FK. 1971. Musculature des membres. In Traité de Zoologie, XVI: 3 (Mammifères) (Grassé PP Ed.). Paris: Masson et Cie. pp. 1–475.

Jouffroy FK, Lessertisseur J. 1959. Reflexions sur les muscles contracteurs des doigts et des orteils (contrahentes digitorum) chez les primates. Ann Sci Nat Zool Ser 12 1: 211–235.

Jouffroy FK, Lessertisseur J. 1960. Les spécialisations anatomiques de la main chez les singes à progression suspendue. Mammalia 24: 93–151.

Kaneff A. 1959. Über die evolution des m. abductor pollicis longus und m. extensor pollicis brevis. Mateil morphol Inst Bulg Akad Wiss 3: 175–196.

Kaneff A. 1968. Zur differenzierung des m. abductor pollicis biventer beim Menschen. Morphol Jahrb 112: 289–303.

Kaneff A. 1969. Umbildung der dorsalen Daumenmuskeln beim Menschen. Verh Anat Ges 63: 625–636.

Kaneff A. 1979. Évolution morphologique des musculi extensores digitorum et abductor pollicis longus chez l'Homme, I, Introduction, méthodologie, M. extensor digitorum. Morphol Jahrb 125: 818–873.

Kaneff A. 1980a. Évolution morphologique des musculi extensores digitorum et abductor pollicis longus chez l'Homme, II, Évolution morphologique des m. extensor digiti minimi, abductor pollicis longus, extensor pollicis brevis et extensor pollicis longus chez l'homme. Morphol Jahrb 126: 594–630.

Kaneff A. 1980b. Évolution morphologique des musculi extensores digitorum et abductor pollicis longus chez l'Homme. III. Évolution morphologique du m. extensor indicis chez l'homme, conclusion générale sur l 'évolution morphologique des musculi extensores digitorum et abductor pollicis longus chez l'homme. Morphol Jahrb 126: 774–815.

Kaneff A. 1986. Die Aufrichtung des Menschen und die mor-phologisches Evolution der Musculi extensores digitorum pedis unter dem Gesichtpunkt der evolutiven Myologie, Teil I. Morph Jahrb 132: 375–419.

Kaneff A, Cihak R. 1970. Modifications in the musculus extensor digitorum lateralis in phylogenesis and in human ontogenesis. Acta Anat Basel 77: 583–604.

Keith A. 1894a. The myology of the Catarrhini: a study in evolution. Unpublished PhD thesis, University of Alberdeen. Aberdeen.

Keith A. 1894b. Notes on a theory to account for the various arrangements of the flexor profundus digitorum in the hand and foot of primates. J Anat Physiol 28: 335–339.

Keith A. 1899. On the chimpanzees and their relationship to the gorilla. Proc Zool Soc Lond 1899: 296–312.

Kleinschmidt A. 1938. Die Schlund-und Kehlorgane des Gorillas "Bobby" unter besonderer—Berücksichtigung der gleichen Organe von Mensch und Orang-Ein Beitrag zur vergleichenden Anatomie des kehlkopfes. Morphol Jahrb 81: 78–157.

Kleinschmidt A. 1950. Zur anatomie des kehlkopfs der Anthropoiden. Anat Anz 97: 367–372.

Kohlbrügge JHF. 1896. Der larynx und die stimmbildung der Quadrumana. Natuurk T Ned Ind 55: 157–175.

Kohlbrügge JHF. 1897. Muskeln und Periphere Nerven der Primaten, mit besonderer Berücksichtigung ihrer Anomalien. Verh K Akad Wet Amsterdam Sec 2 5: 1–246.

Koizumi M, Sakai T. 1995. The nerve supply to coracobrachialis in apes. J Anat 186: 395–403.

Laitman JT. 1977. The ontogenetic and phylogenetic development of the upper respiratory system and basicranium in man. Unpublished PhD thesis, Yale University. Yale.

Lander KF. 1918. The pectoralis minor: a morphological study. J Anat 52: 292–318.

Landsmeer JM. 1984. The human hand in phylogenetic perspective. Bull Hosp Jt Dis Orthop Inst 44: 276–287.

Landsmeer JM. 1986. A comparison of fingers and hand in *Varanus*, opossum and primates. Acta Morphol Neerl Scand 24: 193–221.

Landsmeer JM. 1987. The hand and hominisation. Acta Morphol Neerl Scand. 25: 83–93.

Lessertisseur J. 1958. Doit-on distinguer deux plans de muscles interosseux à la main et au pied des primates? Ann Sci nat Zool Biol Anim 20: 77–103.

Lewis OJ. 1962. The comparative morphology of M. flexor accessorius and the associated long flexor tendons. J Anat 96: 321–333.

Lewis OJ 1964 The evolution of the long flexor muscles of the leg and foot. In International Review of General and Experimental Zoology (Felts WJL, Harrison RJ Eds.). New York: Academic Press. pp. 165–185.

Lewis OJ. 1965. The evolution of the Mm. interossei in the primate hand. Anat Rec 1965 153: 275–287.

Lewis OJ. 1966. The phylogeny of the cruropedal extensor musculature with special reference to the primates. J Anat 100: 865–880.

Lewis OJ. 1989. Functional morphology of the evolving hand and foot.Oxford: Clarendon Press.

Loth E. 1912. Beiträge zur Anthropologie der Negerweichteile (Muskelsystem). Stud Forsch Menschen-u Völkerkunde Stuttgart 9: 1–254.

Loth E. 1931. Anthropologie des parties molles (muscles, intestins, vaisseaux, nerfs peripheriques). Paris: Mianowski-Masson et Cie.

Macalister A. 1873. The muscular anatomy of the gorilla. Proc Royal Irish Acad, Ser 2, 1: 501–506.

Mangini U. 1960. Flexor pollicis longus muscle: its morphology and clinical significance. J Bone Jt Surg 42A: 467–559.

Miller RA 1932. Evolution of the pectoral girdle and forelimb in the primates. Amer J Phys Anthropol 17: 1–56.

Morton DJ. 1922. Evolution of the human foot, Part 1. Am J Phys Anthropol 5: 305–336.

Morton DJ. 1924. The peroneus tertius in gorillas. Anat Rec 17: 323–328.

Owen R. 1859. On the gorilla (*Troglodytes gorilla* Sav.). Proc Zool Soc Lond 1859: 1–23.

Owen R. 1865. Memoir on the gorilla (*Troglodytes gorilla* Savage). London: Taylor & Francis.

Owen R. 1868. The Anatomy of Vertebrates. Vol. 3, Mammals. London: Longmans, Green & Co.

Parsons FG. 1898a. The muscles of mammals, with special relation to human myology, Lecture 1, The skin muscles and muscles of the head and neck. J Anat Physiol 32: 428–450.

Parsons FG. 1898b. The muscles of mammals, with special relation to human myology, a course of lectures delivered at the Royal College of Surgeons of England, Lecture II, The muscles of the shoulder and forelimb. J Anat Physiol 32: 721–752.

Payne RC. 2001. Musculoskeletal adaptations for climbing in hominoids and their role as exaptations for the acquisition of bipedalism. Unpublished PhD thesis, The University of Liverpool, Liverpool.

Pira A. 1913. Beiträge zür Anatomie des Gorilla, I, Das Extremitätenmuskelsystem. Morphol Jahrb 47: 309–354.

Plattner F. 1923. Über die ventral-innervierte und die genuine. rückenmuskulatur bei drei anthropomorphen (*Gorilla gina, Hylobates* und *Troglodytes niger*). Morphol Jb: 241–280.

Preuschoft H. 1961. Muskeln und Gelenke der Hinterextremitat des Gorillas. Morphol Jahrb 101: 432–540.

Preuschoft H. 1963. Muskelgewichte bei gorilla, orangutan und mensch. Anthrop Anz 26: 308–317.

Preuschoft H. 1965. Muskeln and gelenk der vorderextremitat des gorillas. Morph Jb 107: 99–183.

Raven HC. 1950. Regional anatomy of the Gorilla. In The Anatomy of the Gorilla (Gregory WK Ed.). New York: Columbia University Press. pp. 15–188.

Rauwerdink GP. 1993. Muscle fibre and tendon lengths in primate extremities. In Hands of Primates (Preuschoft H, Chivers DJ Eds.). New York: Springer-Verlag. pp. 207–223.

Ruge G. 1885. Über die Gesichtsmuskulatur der halbaffen. Gegen Morph Jahrb 11: 243–315.

Ruge G. 1887a. Untersuchungen uber die Gesichtsmuskeln der Primaten. Leipzig: W. Engelmann.

Ruge G. 1887b. Die vom Facialis innervirten Muskeln des Halses, Nackens und des Schädels einen jungen *Gorilla*. Morphol Jahrb 12: 459–529.

Ruge G. 1897. Über das peripherische gebiet des nervus *facialis* boi wirbelthieren. Leipzig: Festschr f Gegenbaur.

Saban R. 1968. Musculature de la tête. In Traité de Zoologie, XVI, 3 (Mammifères) (Grassé PP Ed.). Paris: Masson et Cie. pp. 229–472.

Sakka M. 1973. Anatomie comparée de l'écaille de l'occipital (squama occipitalis P.N.A.) et des muscles de la nuque chez l'homme et les pongidés, II Partie, Myologie. Mammalia 37: 126–180.

Sakka M. 1977. Ensembles anatomiques cervico-céphaliques: port de tête et évolution des hominides, conséquences phylogéniques sur *Australopithecus*. Mammalia 41: 85–109.

Sarmiento EE. 1994. Terrestrial traits in the hands and feet of gorillas. Am Mus Novit 3091: 1–56.

Schreiber HV. 1934. Zur morphologie der wrimatenhand - rontnenolonische untersuchungen an der handwuriel dei affen. Anat Anz 78: 369–429.

Schreiber HV. 1936. Die extrembewegungen der schimpansenhand, 2, mitteilung zu-zur morphologie der primatehand. Morph Jahrb 77: 22–60.

Schück AC. 1913a. Beiträge zur Myologie der Primaten, I-der m. lat. dorsi und der m. latissimo-tricipitalis. Morphol Jahrb 45: 267–294.

Schück AC. 1913b. Beiträge zur Myologie der Primaten, II-1 die gruppe sterno-cleido-mastoideus, trapezius, omo-cervicalis, 2 die gruppe levator scapulae, rhomboides, serratus anticus. Morphol Jahrb 46: 355–418.

Seiler R. 1970. Differences in the facial musculature of the nasal and upper-lip region in catarrhine primates and man. Z Morphol Anthropol 62: 267–275.

Seiler R. 1971a. A comparison between the facial muscles of Catarrhini with long and short muzzles. Proc 3rd Int Congr Primat Zürich 1970, Vol. l. Basel: Karger. pp. 157–162.

Seiler R. 1971b. Facial musculature and its influence on the facial bones of catarrhine Primates, I. Morphol Jahrb 116: 122–142.

Seiler R. 1971c Facial musculature and its influence on the facial bones of catarrhine Primates, II. Morphol Jahrb 116: 147–185.

Seiler R. 1971d. Facial musculature and its influence on the facial bones of catarrhine Primates, III. Morphol Jahrb 116: 347–376.

Seiler R. 1971e. Facial musculature and its influence on the facial bones of catarrhine Primates, IV. Morphol Jahrb 116: 456–481.

Seiler R. 1976. Die Gesichtsmuskeln. In: Primatologia, Handbuch der Primatenkunde, Bd. 4, Lieferung 6 (Hofer H, Schultz AH, Starck D, Eds.). Basel: Karger. pp. 1–252.

Seiler R. 1977. Morphological and functional differentiation of muscles—studies on the m. frontalis, auricularis superior and auricularis anterior of primates including man. Verh Anat Ges 71: 1385–1388.

Seiler R. 1979a. Criteria of the homology and phylogeny of facial muscles in primates including man, I, Prosimia and Platyrrhina. Morphol Jahrb 125: 191–217.

Seiler R. 1979b. Criteria of the homology and phylogeny of facial muscles in primates including man, II, Catarrhina. Morphol Jahrb 125: 298–323.

Shoshani J, Groves CP, Simons EL, Gunnell F. 1996. Primate phylogeny: morphological vs molecular results. Mol Phylogenet Evol 5: 102–154.

Shrewsbury MM, Marzke MM, Linscheid RL, Reece SP. 2003. Comparative morphology of the pollical distal phalanx. Am J Phys Anthropol 121: 30–47.

Sommer A. 1907. Das Muskelsystem des Gorilla. Jena Z Naturwiss 42: 181–308.

Sonntag CF. 1924. The morphology and evolution of the apes and man. London: John Bale Sons and Danielsson, Ltd.

Starck D, Schneider R. 1960. Respirationsorgane. In: Primatologia III/2 (Hofer H, Schultz AH, Starck D, Eds.). Basel: S. Karger. pp. 423–587.

Steiner PE. 1954. Anatomical observations in a *Gorilla gorilla*. Am J Phys Anthropol 12: 145–165.

Stern JT, Larson SG. 2001. Telemetered electromyography of the supinators and pronators of the forearm in gibbons and chimpanzees: implications for the fundamental positional adaptation of hominoids. Am J Phys Anthopol 115: 253–268.

Stern JT, Wells JP, Jungers WL, Vangor AK, Fleagle JG. 1980a. An electromyographic study of the pectoralis major in atelines and *Hylobates* with special reference to the evolution of a pars clavicularis. Am J Phys Anthropol 52: 13–25.

Stern JT, Wells JP, Jungers WL, Vangor AK. 1980b. An electromyographic study of serratus anterior in atelines and *Alouatta*: implications for hominoid evolution. Am J Phys Anthropol 52: 323–334.

Stewart TD. 1936. The musculature of the anthropoids, I, neck and trunk. Am J Phys Anthropol 21: 141–204.

Straus WL. 1930. The foot musculature of the highland gorilla (*Gorilla beringei*). Q Rev Biol 5: 261–317.

Straus WL. 1941a. The phylogeny of the human forearm extensors. Hum Biol 13: 23–50.

Straus WL. 1941b. The phylogeny of the human forearm extensors (concluded). Hum Biol 13: 203–238.

Straus WL. 1942. Rudimentary digits in primates. Q Rev Biol 17: 228–243.

Susman RL, Nyati L, Jassal MS. 1999. Observations on the pollical palmar interosseus muscle (of Henle). Anat Rec 254: 159–165.

Symington J. 1889. Observations on the myology of the gorilla and chimpanzee. Rep Brit Assoc Adv Sci 59: 629–630.

Testut L. 1883. Le long fléchisseur propre du pouce chez l'homme et les singes. Bull Soc Zool Fr 8: 164–185.

Testut L. 1884. Les anomalies musculaires chez l'homme expliquèes par l'anatomie comparée et leur importance en anthropologie.Paris: Masson.

Tocheri MW, Orr CM, Jacofsky MC, Marzke MW. 2008. The evolutionary history of the hominin hand since the last common ancestor of *Pan* and *Homo*. J Anat 212: 544–562.

Tuttle RH. 1967. Knuckle-walking and the evolution of hominoid hands. Am J Phys Anthrop 26: 171–206.

Tuttle RH. 1969. Quantitative and functional studies on the hands of the Anthropoidea, I, the Hominoidea. J Morphol 128: 309–363.

Tuttle RH. 1970. Postural, propulsive and prehensile capabilities in the cheiridia of chimpanzees and other great apes. In: The Chimpanzee, Vol. 2. (Bourne GH Ed.). Basel: Karger. pp. 167–253.

Tuttle RH. 1972. Relative mass of cheiridial muscles in catarrhine primates. In The Functional and Evolutionary Biology of Primates. (Tuttle RH Ed.). pp. 262–291.

Tuttle RH, Basmajian JV. 1974a. Electromyography of the brachial muscles in *Pan gorilla* and hominoid evolution. Am J Phys Anthropol 41: 71–90

Tuttle RH, Basmajian JV. 1974b. Electromyography of forearm musculature in *Gorilla* and problems related to knuckle-walking. In: Primate Locomotion (Jenkins FA Ed.). New York: Academic Press. pp. 293–347.

Tuttle RH, Basmajian JV. 1976. Electromyography of pongid shoulder muscles and hominoid evolution I—retractors of the humerus and rotators of the scapula. Yearbook Phys Anthropol 20: 491–497.

Tuttle RH, Basmajian JV. 1978a. Electromyography of pongid shoulder muscles II - deltoid, rhomboid and "rotator cuff". Am J Phys Anthropol 49: 47–56.

Tuttle RH, Basmajian JV. 1978b. Electromyography of pongid shoulder muscles III—quadrupedal positional behavior. Am J Phys Anthropol 49: 57–70.

Tuttle RH, Cortright G. 1988. Positional behavior, adaptive complexes, and evolution. In Orang-Utan Biology (Schwartz JH Ed.). Oxford: Oxford University Press, pp. 311–330.

Tuttle RH, Hollowed JR, Basmajian JV. 1992. Electromyography of pronators and supinators in great apes. Am J Phys Anthropol 87: 215–26.

Vrolik W. 1841. Recherches d' anatomie comparé, sur le chimpanzé. Amsterdam: Johannes Miller.

Yoshikawa T. 1961. The lamination of the m. masseter of the crab-eating monkey, orang-utan and gorilla. Primates 3: 81.

Weidenreich YF. 1951. The morphology of Solo man, 3. Anthropol Pap Am Mus Nat hist 43: 205–290.

Whitehead PF. 1993. Aspects of the anthropoid wrist and hand. In Postcranial Adaptation in Nonhuman Primates (Gebo DL Ed.). DeKalb: Northern Illinois University Press. pp. 96–120.

Wilkinson JL. 1953. The insertions of the flexores pollicis longus et digitorum profundus. J Anat 87: 75–88.

Wislocki GB. 1932. On the female reproductive tract of the gorilla with a comparison with that of other primates. Contributions Embryol 23: 163–204.

Wood Jones F. 1920. The principles of anatomy as seen in the hand.London: J. & A Churchill.

Zihlman AL. 2000. Body mass in lowland gorillas: a quantitative analysis. Am J Phys Anthropol 113: 61–78.

Appendix II
Literature Cited, not Including Information about the Muscles of Gorillas

Diogo R. 2004a. Morphological evolution, aptations, homoplasies, constraints, and evolutionary trends: catfishes as a case study on general phylogeny and macroevolution. Enfield: Science Publishers.

Diogo R. 2004b. Muscles versus bones: catfishes as a case study for an analysis on the contribution of myological and osteological structures in phylogenetic reconstructions. Anim Biol 54: 373–391.

Diogo R. 2007. On the origin and evolution of higher-clades: osteology, myology, phylogeny and macroevolution of bony fishes and the rise of tetrapods. Enfield: Science Publishers.

Harrison DFN. 1995. The anatomy and physiology of the mammalian larynx. Cambridge: University Press.

Kelemen G. 1948.The anatomical basis of phonation in the chimpanzee. J Morphol 82: 229–256.

Kelemen G. 1969. Anatomy of the larynx and the anatomical basis of vocal performance. In: The Chimpanzee—Vol. 1—Anatomy, behaviour and diseases of chimpanzees (Bourne GH Ed.). Basel: Karger. pp. 35–186.

Köntges G, Lumsden A. 1996. Rhombencephalic neural crest segmentation is preserved throughout craniofacial ontogeny. Development 122: 3229–3242.

Lightoller GS. 1928. The facial muscles of three orang utans and two ceropithecidae. J Anat 63: 19–81.

Lightoller GS. 1934. The facial musculature of some lesser primates and a *Tupaia*. Proc Zool Soc Lond 1934: 259–309.

Lightoller GS. 1939. Probable homologues, a study of the comparative anatomy of the mandibular and hyoid arches and their musculature—Part I—comparative myology. Trans Zool Soc Lond 24: 349–382.

Netter FH. 2006. Atlas of human anatomy. 4th ed. Philadelphia: Saunders.

Rajalakshmi R, Anu R, Soubhagya N, Rajanigandha V, Pai M, Ashwin K. 2008. A study of anatomical variability of the omohyoid muscle and its clinical relevance. Clinics 63: 521–524.

Swindler DR, Wood CD. 1973. An atlas of primate gross anatomy: baboon, chimpanzee and men. Seattle: University of Washington Press.

Terminologia Anatomica. 1998. Federative Committee on Anatomical Terminology. Stuttgard: Georg Thieme.

Index

About The Authors

Rui Diogo is a biologist undertaking research at the Center for the Advanced Study of Hominid Paleobiology of George Washington University (US). He has numerous publications, and is one of the editors of the books *Catfishes* and *Gonorynchiformes and ostariophysan interrelationships—a comprehensive review*. He is the single author of the books *Morphological evolution, aptations, homoplasies, constraints and evolutionary trends—catfishes as a case study on general phylogeny and macroevolution* and *The origin of higher clades—osteology, myology, phylogeny and evolution of bony fishes and the rise of tetrapods*, and the first author of the book *Muscles of vertebrates—comparative anatomy, evolution, homologies and development*.

Josep Potau is Professor at the Department of Anatomy and Embryology of the University of Barcelona (Spain), and is the director of the University's Center for the Study of Comparative and Evolutionary Anatomy. His current research focuses on the analysis of functional and anatomical adaptations associated with the evolution of different types of locomotion and of the upper limb musculature within primates. He has published many papers and book chapters on functional and comparative anatomy.

Juan Pastor is Professor at the Department of Anatomy of the University of Valladolid (Spain), and is the director of the University's Anatomical Museum, which has the largest osteological collection of Spain. He has published many papers on comparative anatomy and anthropology.

Félix de Paz is Professor at the Department of Anatomy of the University of Valladolid, and is a member of the Royal Academy of Medicine and Surgery of Valladolid (Spain). He has published many papers on comparative anatomy and anthropology.

Mercedes Barbosa is Professor at the Department of Anatomy of the University of Valladolid (Spain), and is a member of the Anatomical Society of Spain. She has published many papers on physical anthropology.

Eva Ferrero is a biologist and graduated from the University of León (Spain). She is undertaking research at the Department of Anatomy and the Anatomical Museum of the University of Valladolid (Spain), on the comparative anatomy of primates and other mammals.

Gaëlle Bello is a biologist and graduated from the University of La Coruña (Spain). She completed her Masters in Primatology at the University of Barcelona (Spain), and is now undertaking a PhD at the University of Barcelona on the evolution of the scapula within primates and its adaptations to different types of locomotion.

Bernard Wood is University Professor of Human Origins and directs the Center for the Advanced Study of Hominid Paleobiology of George Washington University (USA). His edited publications include *Food Acquisition and Processing in Primates* and *Major Topics in Primate and Human Evolution*. He is author or co-author of *The Evolution of Early Man, Human Evolution, Koobi Fora Research Project—Hominid Cranial Remains (Vol. 4), Human Evolution—A Very Short Introduction*, and he is the editor of the forthcoming *Wiley-Blackwell Encyclopedia of Human Evolution*.

Color Plate Section

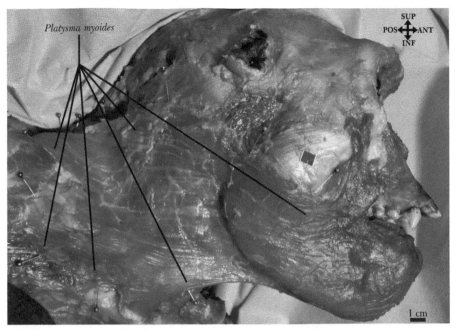

Fig. 1 *Gorilla gorilla* (VU GG1, adult female): lateral view of the right facial musculature, showing the platysma myoides. Note that there is no well developed platysma cervicale going to the nuchal region. In this figure and the remaining figures of the atlas, the names of the muscles are in italics, and SUP, INF, ANT, POS, MED, LAT, VEN, DOR, PRO and DIS refer to superior, inferior, anterior, posterior, medial, lateral, ventral, dorsal, proximal and distal, respectively.

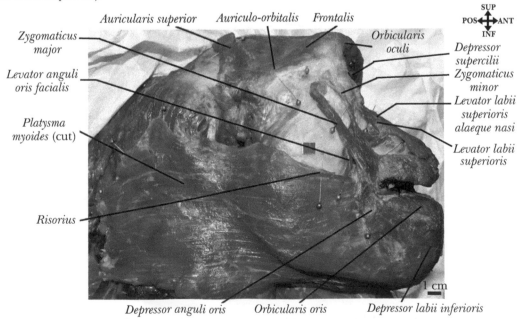

Fig. 2 *Gorilla gorilla* (VU GG1, adult female): lateral view of the right facial musculature, showing a thin muscular structure that corresponds to the risorius of modern humans.

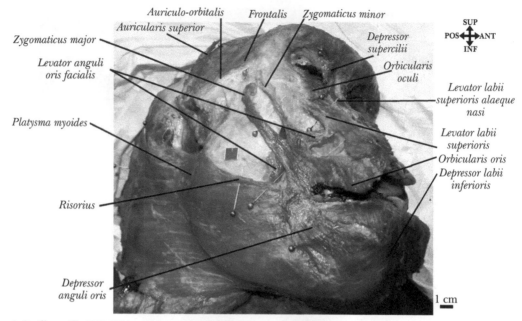

Fig. 3 *Gorilla gorilla* (VU GG1, adult female): fronto-lateral view of the right facial musculature, showing a thin muscular structure that corresponds to the risorius of modern humans.

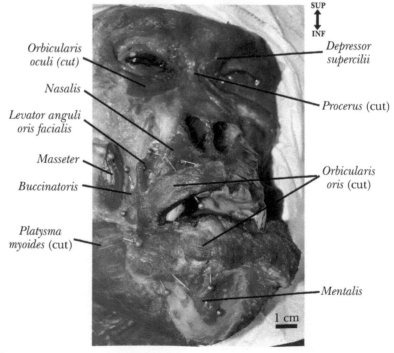

Fig. 4 *Gorilla gorilla* (VU GG1, adult female): frontal view of the deep facial musculature, after removing the levator labii superioris, levator labii superioris alaeque nasi, depressor labii inferioris, depressor anguli oris, zygomaticus major, and zygomaticus minor and after partially cutting the orbicularis oris and orbicularis oculi.

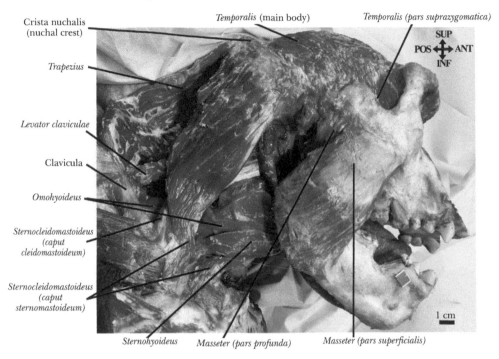

Fig. 5 *Gorilla gorilla* (VU GG1, adult female): lateral view of the right head and neck muscles, as well as some pectoral and upper limb muscles such as the levator claviculae and pectoralis major.

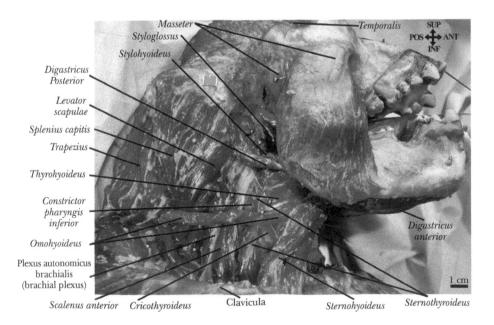

Fig. 6 *Gorilla gorilla* (VU GG1, adult female): lateral view of the right head and neck muscles, after removal of levator claviculae and sternocleidomastoideus.

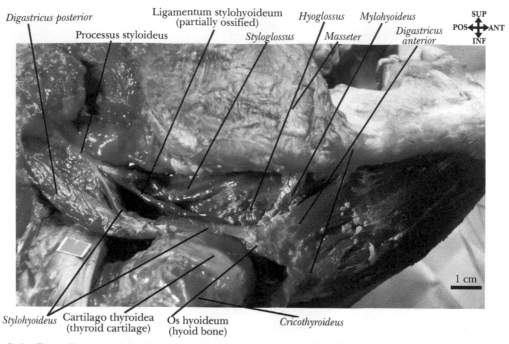

Fig. 7 *Gorilla gorilla* (VU GG1, adult female): latero-ventral view of the right tongue and suprahyoid muscles; note the ossification of the stylohyoid ligament, and also how the intermediate tendon of the digastric pierces the stylohyoideus.

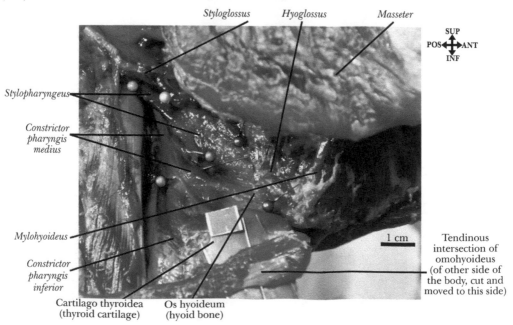

Fig. 8 *Gorilla gorilla* (VU GG1, adult female): latero-ventral view of the right tongue and pharyngeal muscles after removal of all the infrahyoid and suprahyoid muscles except the mylohyoideus and geniohyoideus; the omohyoideus of the other side of the body is also shown—note the small tendinous intersection of this latter muscle.

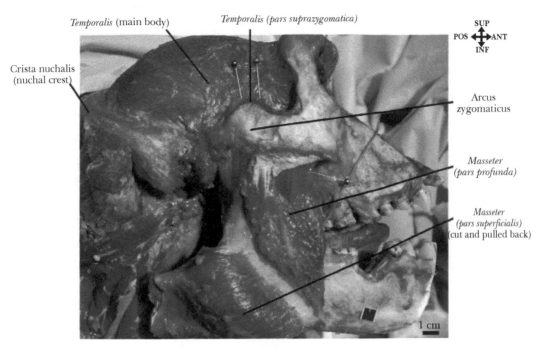

Temporalis (main body)   Temporalis (pars suprazygomatica)

SUP
POS ⬌ ANT
INF

Crista nuchalis
(nuchal crest)

Arcus
zygomaticus

Masseter
(pars profunda)

Masseter
(pars superficialis)
(cut and pulled back)

1 cm

Fig. 9 *Gorilla gorilla* (VU GG1, adult female): lateral view of the right masseter and temporalis. Note the small pars suprazygomatica of the temporalis, which is mainly covered by the zygomatic arch.

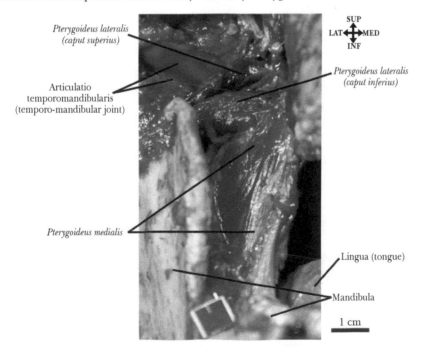

Pterygoideus lateralis
(caput superius)

SUP
LAT ⬌ MED
INF

Pterygoideus lateralis
(caput inferius)

Articulatio
temporomandibularis
(temporo-mandibular joint)

Pterygoideus medialis

Lingua (tongue)

Mandibula

1 cm

Fig. 10 *Gorilla gorilla* (VU GG1, adult female): frontal view of the right pterygoideus lateralis and pterygoideus medialis.

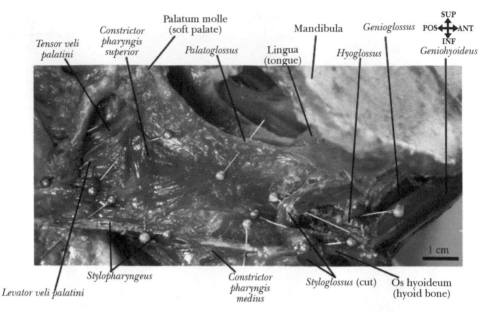

Fig. 11 *Gorilla gorilla* (VU GG1, adult female): lateral view of the right pharyngeal and tongue muscles after removal of all suprahyoid and infrahyoid muscles except the geniohyoideus; note the presence of a few muscular fibers within the palatoglossal fold, which seem to correspond to the fibers of the palatoglossus muscle of humans.

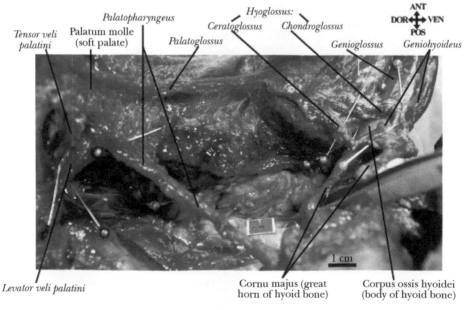

Fig. 12 *Gorilla gorilla* (VU GG1, adult female): lateral view of the right pharyngeal and tongue muscles after removal of all suprahyoid and infrahyoid muscles except the geniohyoideus. The styloglossus and the superior and middle constrictors of the pharynx were also removed. Note the division of the hyoglossus into a ceratoglossus and a chondroglossus.

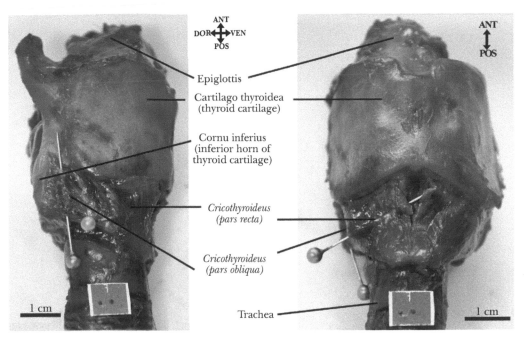

Fig. 13 *Gorilla gorilla* (VU GG1, adult female): lateral (on the left) and ventral (on the right) view of the laryngeal muscles.

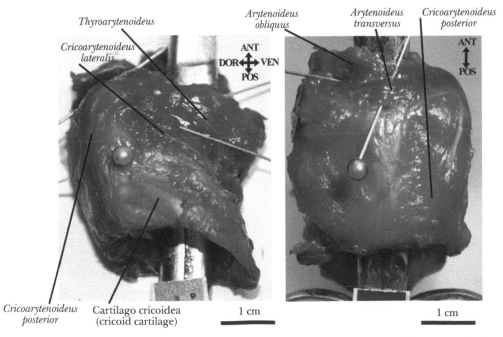

Fig. 14 *Gorilla gorilla* (VU GG1, adult female): lateral (on the left) and dorsal (on the right) views of the laryngeal muscles after removal of the cricothyroideus, trachea and thyroid cartilage.

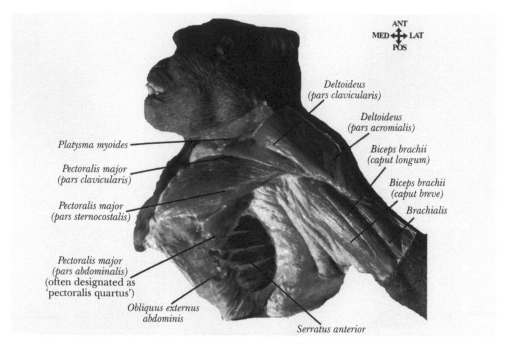

Fig. 15 *Gorilla gorilla* (VU GG1, adult female): ventrolateral view of the left pectoral and arm muscles.

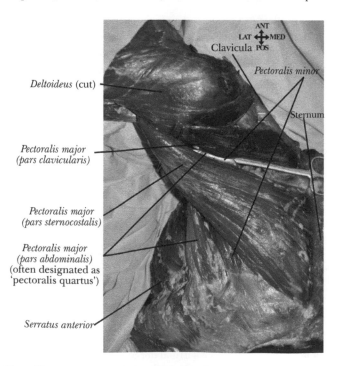

Fig. 16 *Gorilla gorilla* (VU GG1, adult female): ventral view of the right serratus anterior, deltoideus, pectoralis minor and pectoralis major.

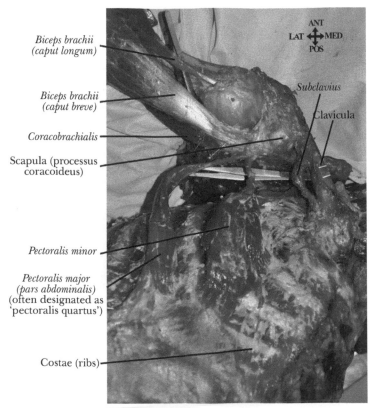

Biceps brachii
(caput longum)

Biceps brachii
(caput breve)

Coracobrachialis

Scapula (processus
coracoideus)

Pectoralis minor

Pectoralis major
(pars abdominalis)
(often designated as
'pectoralis quartus')

Costae (ribs)

Subclavius

Clavicula

Fig. 17 *Gorilla gorilla* (VU GG1, adult female): ventral view of the right pectoralis minor, subclavius, biceps brachii, coracobrachialis and pars abdominalis of the pectoralis major after removing the deltoideus and the pars clavicularis and pars sternocostalis of the pectoralis major.

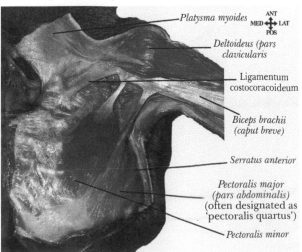

Platysma myoides

Deltoideus (pars
clavicularis

Ligamentum
costocoracoideum

Biceps brachii
(caput breve)

Serratus anterior

Pectoralis major
(pars abdominalis)
(often designated as
'pectoralis quartus')

Pectoralis minor

Fig. 18 *Gorilla gorilla* (VU GG2, adult female): ventrolateral view of the left pectoral and arm muscles after removal of the pars clavicularis and pars sternocostalis of the pectoralis major; note the connection between the pars abdominalis of the pectoralis major and the biceps brachii.

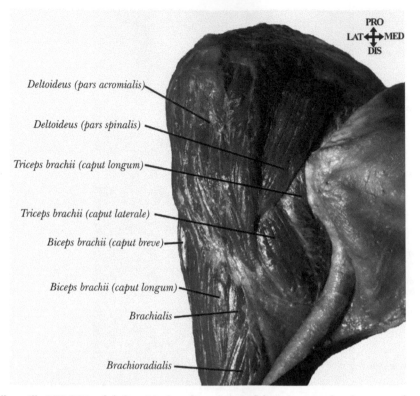

Deltoideus (pars acromialis)

Deltoideus (pars spinalis)

Triceps brachii (caput longum)

Triceps brachii (caput laterale)

Biceps brachii (caput breve)

Biceps brachii (caput longum)

Brachialis

Brachioradialis

PRO
LAT — MED
DIS

Fig. 19 *Gorilla gorilla* (VU GG1, adult female): dorsolateral view of the left pectoral and arm muscles.

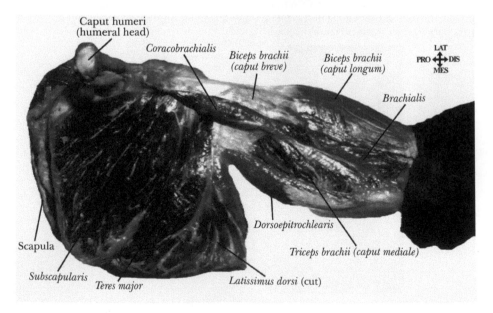

Caput humeri
(humeral head)

Coracobrachialis

Biceps brachii
(caput breve)

Biceps brachii
(caput longum)

Brachialis

LAT
PRO — DIS
MES

Dorsoepitrochlearis

Triceps brachii (caput mediale)

Scapula

Subscapularis

Teres major

Latissimus dorsi (cut)

Fig. 20 *Gorilla gorilla* (VU GG1, adult female): ventral view of the left pectoral and arm muscles after removal of deltoideus.

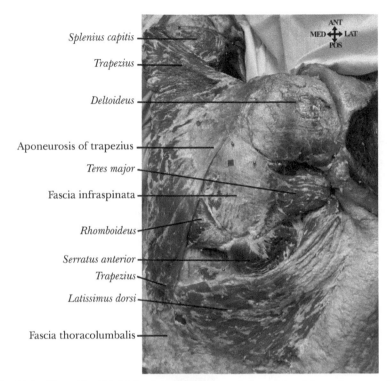

Splenius capitis

Trapezius

Deltoideus

Aponeurosis of trapezius

Teres major

Fascia infraspinata

Rhomboideus

Serratus anterior

Trapezius

Latissimus dorsi

Fascia thoracolumbalis

Fig. 21 *Gorilla gorilla* (VU GG1, adult female): dorsal view of the right neck, pectoral and arm muscles.

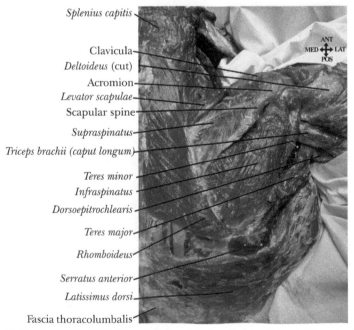

Splenius capitis

Clavicula

Deltoideus (cut)

Acromion

Levator scapulae

Scapular spine

Supraspinatus

Triceps brachii (caput longum)

Teres minor

Infraspinatus

Dorsoepitrochlearis

Teres major

Rhomboideus

Serratus anterior

Latissimus dorsi

Fascia thoracolumbalis

Fig. 22 *Gorilla gorilla* (VU GG1, adult female): dorsal view of the right pectoral and arm muscles after removing the trapezius and cutting the deltoideus in order to show the infraspinatus, supraspinatus and teres minor.

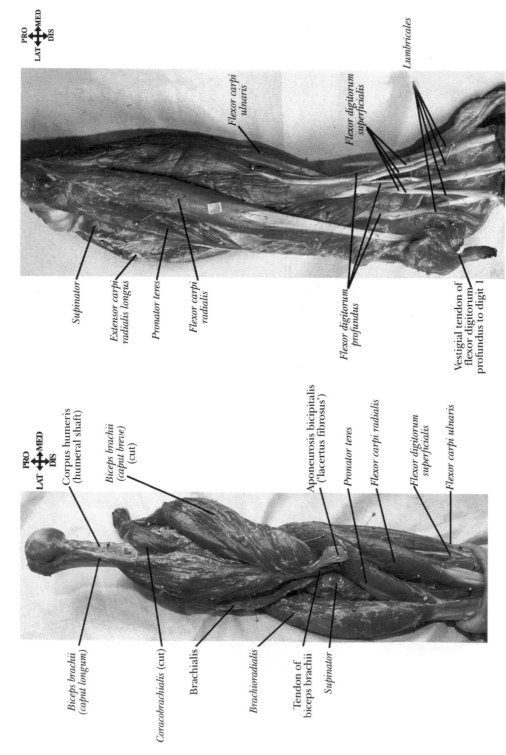

Fig. 23 *Gorilla gorilla* (VU GG1, adult female): ventral view of the right arm and forearm muscles. Note the absence of the palmaris longus.

Fig. 24 *Gorilla gorilla* (VU GG1, adult female): ventral view of the right forearm and hand muscles. Note the absence of the palmaris longus, and of a distinct flexor pollicis longus.

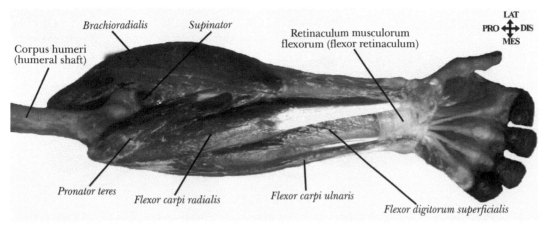

Fig. 25 *Gorilla gorilla* (VU GG1, adult female): ventral view of the left forearm muscles.

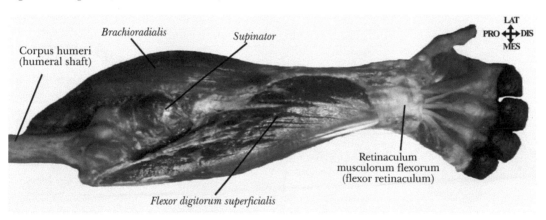

Fig. 26 *Gorilla gorilla* (VU GG1, adult female): ventral view of the left forearm muscles after removal of pronator teres, flexor carpi radialis and flexor carpi ulnaris.

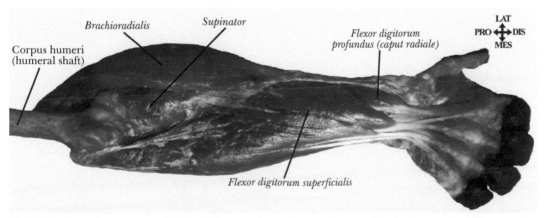

Fig. 27 *Gorilla gorilla* (VU GG1, adult female): same view as in Fig. 26, except that the flexor retinaculum was also removed. Note there is no tendon running from the main tendon of the caput radiale of the flexor digitorum profundus to digit 1.

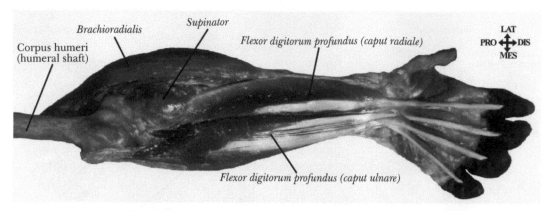

Fig. 28 *Gorilla gorilla* (VU GG1, adult female): same view as in Fig. 27, except that the flexor digitorum superficialis was also removed. Note there is no tendon running from the main tendon of the caput radiale of the flexor digitorum profundus to digit 1.

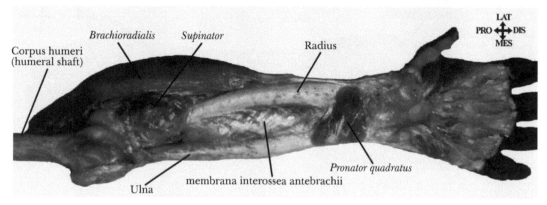

Fig. 29 *Gorilla gorilla* (VU GG1, adult female): same view as in Fig. 28, except that the flexor digitorum profundus was also removed.

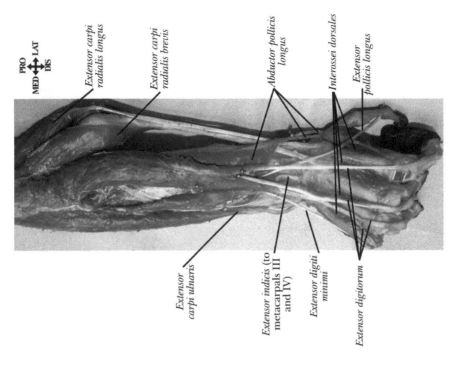

PRO
MED ⟷ LAT
DIS

Extensor carpi
radialis longus

Extensor carpi
radialis brevis

Abductor pollicis
longus

Interossei dorsales

Extensor
pollicis longus

Extensor
carpi ulnaris

Extensor indicis (to
metacarpals III
and IV)

Extensor digiti
minimi

Extensor digitorum

Fig. 31 *Gorilla gorilla* (VU GG1, adult female): dorsal view of the right forearm and hand muscles. Note the extensor indicis insertion onto the base of metacarpals III and IV, and the absence of a separate extensor pollicis brevis.

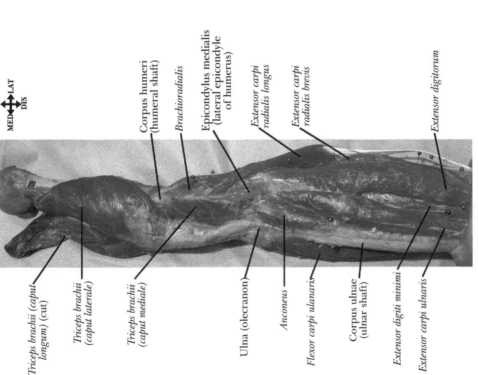

PRO
MED ⟷ LAT
DIS

Corpus humeri
(humeral shaft)

Brachioradialis

Epicondylus medialis
(lateral epicondyle
of humerus)

Extensor carpi
radialis longus

Extensor carpi
radialis brevis

Extensor digitorum

Triceps brachii (caput
longum) (cut)

Triceps brachii
(caput laterale)

Triceps brachii
(caput mediale)

Ulna (olecranon)

Ancoeus

Flexor carpi ulnaris

Corpus ulnae
(ulnar shaft)

Extensor digiti minimi

Extensor carpi ulnaris

Fig. 30 *Gorilla gorilla* (VU GG1, adult female): dorsal view of the right arm and forearm muscles. Note the absence of the epitrochleoanconeus.

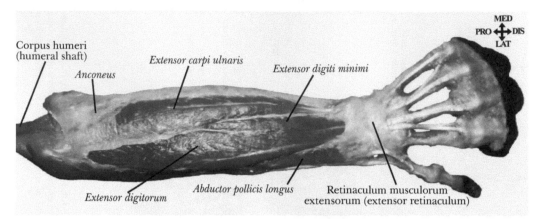

Fig. 32 *Gorilla gorilla* (VU GG1, adult female): dorsal view of the left forearm muscles after removal of the hand musculature.

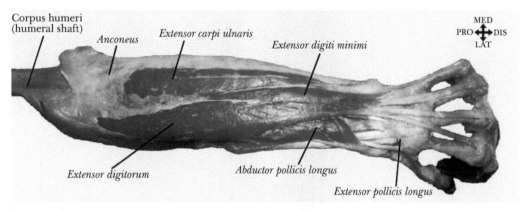

Fig. 33 *Gorilla gorilla* (VU GG1, adult female): same view as in Fig. 32, except that the extensor retinaculum was also removed. Note the abductor pollicis longus has a single, undivided fleshy belly.

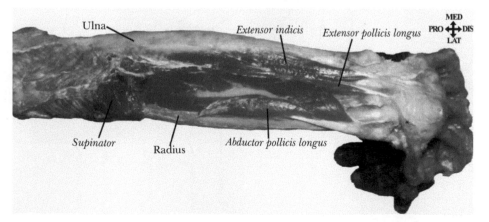

Fig. 34 *Gorilla gorilla* (VU GG1, adult female): same view as in Fig. 33, except that the extensor digitorum, extensor digiti minimi and extensor carpi ulnaris were also removed. Note the abductor pollicis longus has a single, undivided fleshy belly.

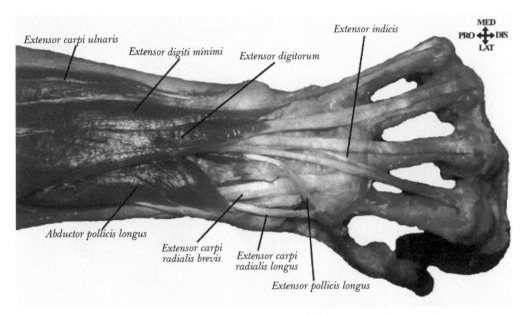

Fig. 35 *Gorilla gorilla* (VU GG1, adult female): dorsal view of the left forearm muscles; note that, contrary to the condition found on the other side of the body, in which the extensor indicis goes exclusively to the base of metacarpals III and IV (shown in Fig. 31), on this side of the body this muscle goes to the bases of metacarpals III and IV (shown in Fig. 36) but also to the middle and distal phalanges of digit 2 (shown here and in Fig. 36).

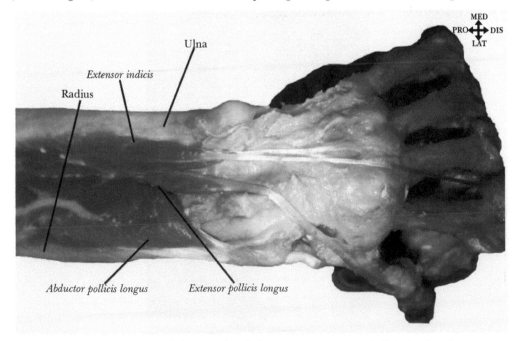

Fig. 36 *Gorilla gorilla* (VU GG1, adult female): dorsal view of the left extensor indicis showing that, contrary to the condition found on the other side of the body, in which the extensor indicis goes exclusively to the base of metacarpals III and IV (shown in Fig. 31), on this side of the body this muscle goes to the bases of metacarpals III and IV (shown here) but also to the middle and distal phalanges of digit 2 (shown here and in Fig. 35).

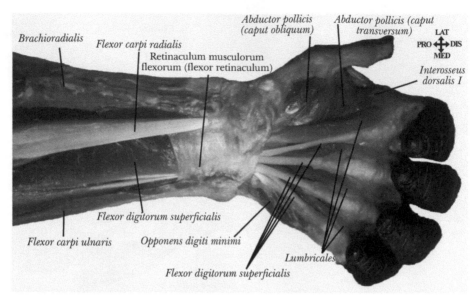

Brachioradialis

Flexor carpi radialis

Retinaculum musculorum
flexorum (flexor retinaculum)

*Abductor pollicis
(caput obliquum)*

*Abductor pollicis (caput
transversum)*

LAT
PRO ↔ DIS
MED

*Interosseus
dorsalis I*

Flexor digitorum superficialis

Flexor carpi ulnaris   *Opponens digiti minimi*

Lumbricales

Flexor digitorum superficialis

Fig. 37a *Gorilla gorilla* (VU GG1, adult female): ventral view of the muscles of the left forearm and hand after removal of abductor pollicis brevis, flexor pollicis brevis, opponens pollicis, palmaris brevis, abductor digiti minimi and flexor digiti minimi brevis.

Extensor digitorum

DIS
MED ↔ LAT
PRO

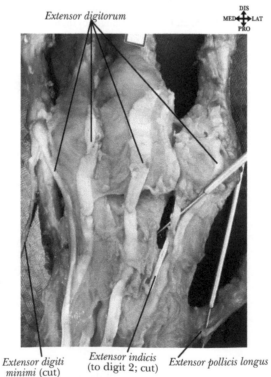

*Extensor digiti
minimi* (cut)

*Extensor indicis
(to digit 2; cut)*

Extensor pollicis longus

Fig. 37b *Gorilla gorilla* (CMS GG1, adult male): dorsal view of the left forearm muscles. Note that, contrary to the condition found in the specimen VU GG1 (see Figs. 31, 35 and 36), in this specimen the extensor indicis goes exclusively to digit 2.

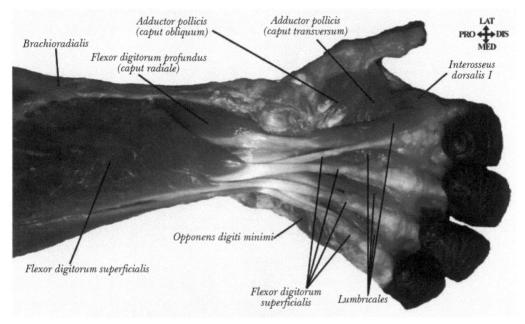

Fig. 38a *Gorilla gorilla* (VU GG1, adult female): same view as in Fig. 37a, with the flexor carpi radialis, flexor carpi ulnaris and flexor retinaculum also removed.

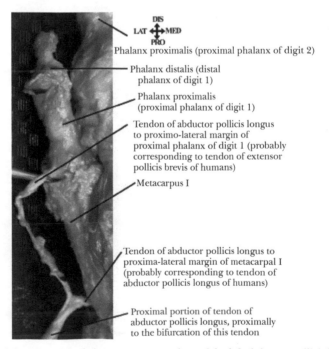

Fig. 38b *Gorilla gorilla* (CMS GG1, adult male): dorsal view of the two tendons of the left abductor pollicis longus, one going to the proximo-lateral margin of the proximal phalanx of digit 1 (thus probably corresponding to the tendon of the extensor pollicis brevis of modern humans), the other going to the proximo-lateral margin of meta-carpal I (thus probably corresponding to the tendon of the abductor pollicis longus of modern humans).

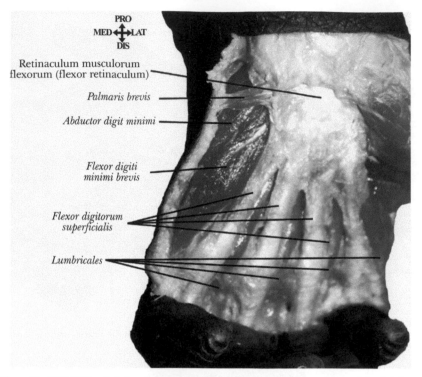

PRO
MED ✛ LAT
DIS

Retinaculum musculorum
flexorum (flexor retinaculum)

Palmaris brevis

Abductor digit minimi

*Flexor digiti
minimi brevis*

*Flexor digitorum
superficialis*

Lumbricales

Fig. 39 *Gorilla gorilla* (VU GG1, adult female): ventral view of the hypothenar muscles of the left hand.

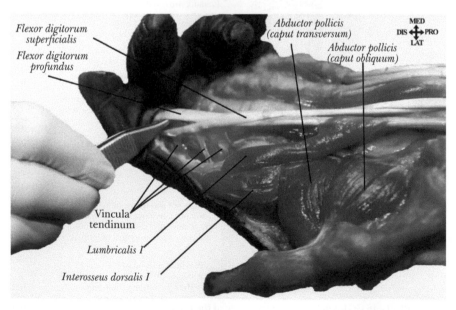

*Flexor digitorum
superficialis*

*Flexor digitorum
profundus*

*Abductor pollicis
(caput transversum)*

*Abductor pollicis
(caput obliquum)*

MED
DIS ✛ PRO
LAT

Vincula
tendinum

Lumbricalis I

Interosseus dorsalis I

Fig. 40 *Gorilla gorilla* (VU GG1, adult female): ventrolateral view of the thenar muscles of the left hand after removal of abductor pollicis brevis, opponens pollicis, and flexor pollicis brevis. Note the relation between the tendons of the flexor digitorum profundus and of the flexor digitorum superficialis.

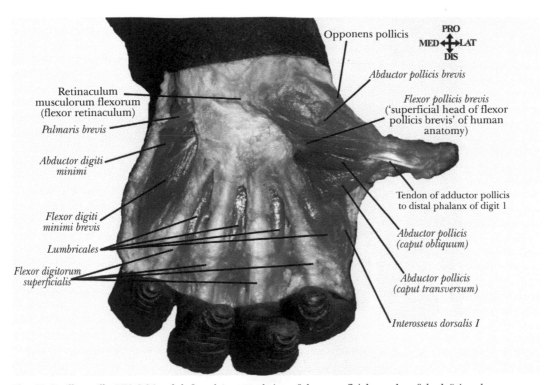

Fig. 41 *Gorilla gorilla* (VU GG1, adult female): ventral view of the superficial muscles of the left hand.

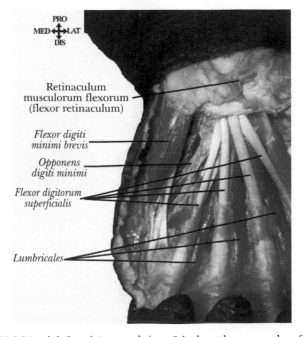

Fig. 42 *Gorilla gorilla* (VU GG1, adult female): ventral view of the hypothenar muscles of the left hand after removal of palmaris brevis and abductor digiti minimi.

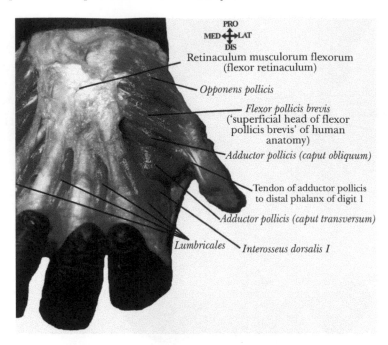

PRO
MED◆◆LAT
DIS

Retinaculum musculorum flexorum
(flexor retinaculum)

Opponens pollicis

Flexor pollicis brevis
('superficial head of flexor
pollicis brevis' of human
anatomy)

Adductor pollicis (caput obliquum)

Tendon of adductor pollicis
to distal phalanx of digit 1

Adductor pollicis (caput transversum)

Lumbricales   *Interosseus dorsalis I*

Fig. 43 *Gorilla gorilla* (VU GG1, adult female): ventral view of the thenar muscles of the left hand after removal of abductor pollicis brevis.

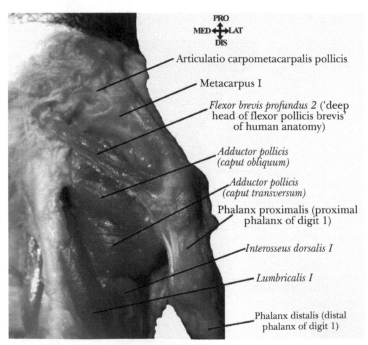

PRO
MED◆◆LAT
DIS

Articulatio carpometacarpalis pollicis

Metacarpus I

Flexor brevis profundus 2 ('deep
head of flexor pollicis brevis'
of human anatomy)

*Adductor pollicis
(caput obliquum)*

*Adductor pollicis
(caput transversum)*

Phalanx proximalis (proximal
phalanx of digit 1)

Interosseus dorsalis I

Lumbricalis I

Phalanx distalis (distal
phalanx of digit 1)

Fig. 44 *Gorilla gorilla* (VU GG1, adult female): ventral view of the thenar muscles of the left hand after the removal of abductor pollicis brevis, of the opponens pollicis and of the 'superficial head of the flexor pollicis brevis' of human anatomy.

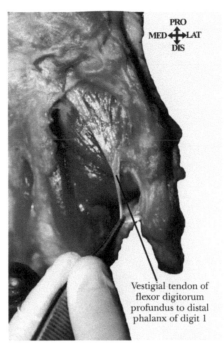

Fig. 45 *Gorilla gorilla* (VU GG1, adult female): ventral view of the vestigial tendon of the flexor digitorum profundus to digit 1, which runs from fascia over the adductor pollicis to the distal phalanx of the left thumb.

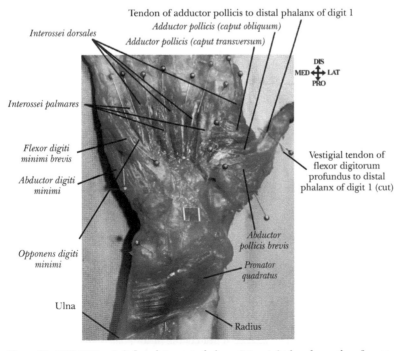

Fig. 46 *Gorilla gorilla* (VU GG1, adult female): ventral view of the right hand muscles after removal of the lumbricales and all the forearm muscles except the pronator quadratus.

Phalanx distalis (distal phalanx of digit 1)

DIS
MED ← → LAT
PRO

Vestigial tendon of flexor digitorum profundus to distal phalanx of digit 1 (cut)

Phalanx proximalis (proximal phalanx of digit 1)

Tendon of adductor pollicis to distal phalanx of digit 1

Adductor pollicis (caput transversum)

Adductor pollicis (caput obliquum)

Vestigial tendon of flexor digitorum profundus to distal phalanx of digit 1 (cut)

TDAS-AD (thin deep additional slip of adductor pollicis; probably corresponds to 'interosseous volaris primus of Henle' of human anatomy)

Metacarpus I

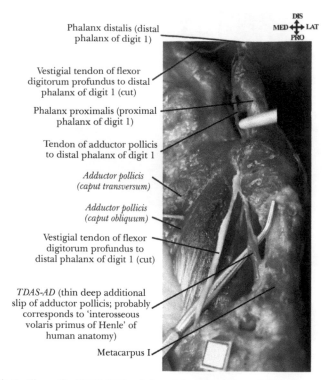

Fig. 47 *Gorilla gorilla* (CMS GG1, adult male): ventral view of the oblique and transverse heads of the right adductor pollicis, as well as of the thin deep additional slip (TDAS-AD) of this muscle, which probably corresponds to the 'interosseous volaris primus of Henle' of modern human anatomy.

Flexor digiti minimi brevis (cut)

Interosseous dorsalis 1

DIS
MED ← → LAT
PRO

Opponens digiti minimi

Flexor brevis profundus ('deep head of flexor pollicis brevis' of human anatomy)

Interosseous palmaris 3

Phalanx proximalis (proximal phalanx of digit 1)

Opponens pollicis

Abductor digiti minimi (cut)

Abductor pollicis brevis (cut)

Flexor pollicis brevis ('superficial head of flexor pollicis brevis' of human anatomy; cut)

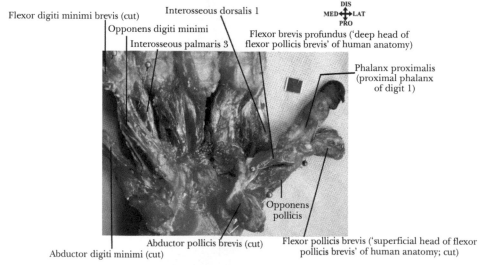

Fig. 48 *Gorilla gorilla* (VU GG1, adult female): ventral view of the right hand muscles after removing the lumbricales and adductor pollicis and cutting the 'superficial head of the flexor pollicis brevis of modern human anatomy', the flexor digiti minimi brevis, the abductor pollicis brevis, and the abductor digiti minimi. Note the well-developed flexor brevis profundus 2 ('deep head of the flexor pollicis brevis' of modern human anatomy).

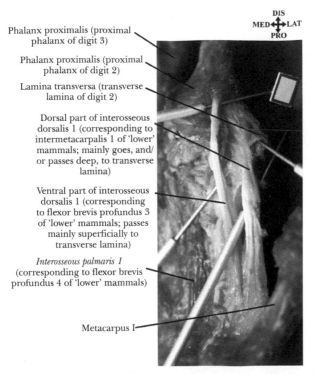

Phalanx proximalis (proximal phalanx of digit 3)

Phalanx proximalis (proximal phalanx of digit 2)

Lamina transversa (transverse lamina of digit 2)

Dorsal part of interosseous dorsalis 1 (corresponding to intermetacarpalis 1 of 'lower' mammals; mainly goes, and/ or passes deep, to transverse lamina)

Ventral part of interosseous dorsalis 1 (corresponding to flexor brevis profundus 3 of 'lower' mammals; passes mainly superficially to transverse lamina)

Interosseous palmaris 1 (corresponding to flexor brevis profundus 4 of 'lower' mammals)

Metacarpus I

Fig. 49 *Gorilla gorilla* (CMS GG1, adult male): ventral view of the ventral and dorsal portions of the right interosseous dorsalis 1, corresponding respectively to the flexor brevis profundus 3 and to the intermetacarpalis 1 of 'lower' mammals. As is usually the case in modern humans and some other primate taxa (but not in chimpanzees), in gorillas the flexores breves profundi 3, 5, 6 and 8 are associated with the intermetacarpales 1, 2, 3 and 4 in order to form the interossei dorsales 1, 2, 3 and 4, respectively.

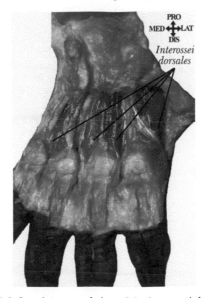

Interossei dorsales

Fig. 50 *Gorilla gorilla* (VU GG1, adult female): ventral view of the interossei dorsales of the left hand.

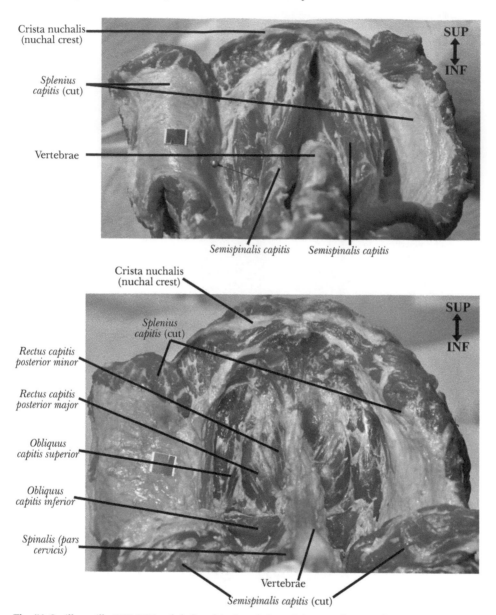

Fig. 51 *Gorilla gorilla* (VU GG1, adult female): dorsal view of the anterior muscles of the back; in the photograph on the bottom the semispinalis capitis was cut in order to show the deep musculature.

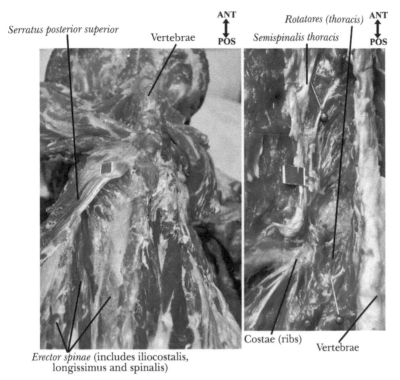

Serratus posterior superior Vertebrae ANT ↕ POS *Rotatores (thoracis)* ANT ↕ POS *Semispinalis thoracis*

Costae (ribs) Vertebrae

Erector spinae (includes iliocostalis,
longissimus and spinalis)

Fig. 52 *Gorilla gorilla* (VU GG1, adult female): the main figure (on the left) is a dorsal view of the back muscles, after reflecting the splenius capitis, splenius cervicis, semispinalis capitis and semispinalis cervicis, in order to show the deep musculature. The figure on the right shows the semispinalis thoracis and the rotatores (thoracis).

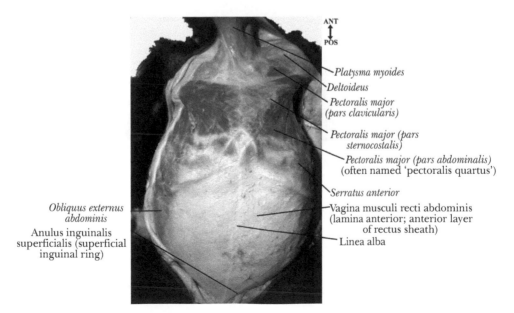

ANT ↕ POS

Platysma myoides

Deltoideus

*Pectoralis major
(pars clavicularis)*

*Pectoralis major (pars
sternocostalis)*

Pectoralis major (pars abdominalis)
(often named 'pectoralis quartus')

Serratus anterior

Vagina musculi recti abdominis
(lamina anterior; anterior layer
of rectus sheath)

Linea alba

*Obliquus externus
abdominis*

Anulus inguinalis
superficialis (superficial
inguinal ring)

Fig. 53 *Gorilla gorilla* (VU GG3, adult male): ventral view of the pectoral and abdominal musculature.

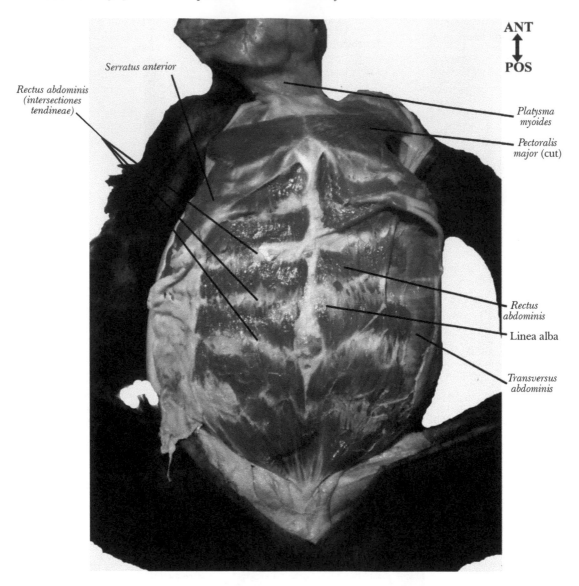

Fig. 54 *Gorilla gorilla* (VU GG3, adult male): ventral view of the pectoral and abdominal musculature after removing the anterior layer of the rectus sheath and part of the pectoralis major and of the obliquus externus abdominis.

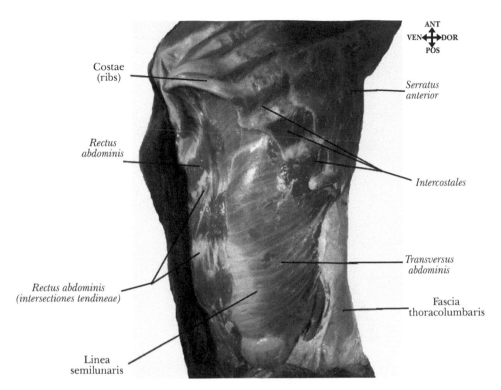

Fig. 55 *Gorilla gorilla* (VU GG3, adult male): ventrolateral view of the pectoral and left abdominal musculature after removing the anterior layer of the rectus sheath and part of the pectoralis major and of the obliquus externus abdominis.

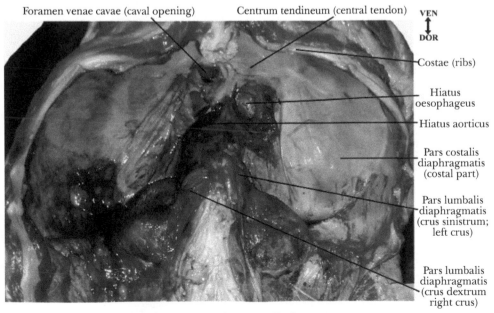

Fig. 56 *Gorilla gorilla* (VU GG3, adult male): posterior view of the diaphragma.

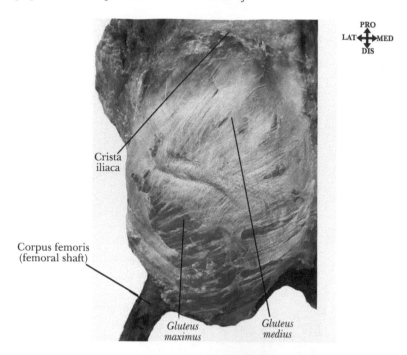

Fig. 57 *Gorilla gorilla* (VU GG1, adult female): dorsal view of the left buttock musculature.

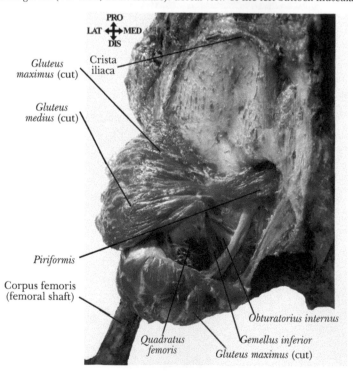

Fig. 58 *Gorilla gorilla* (VU GG1, adult female): dorsal view of the deep muscles of the left buttock.

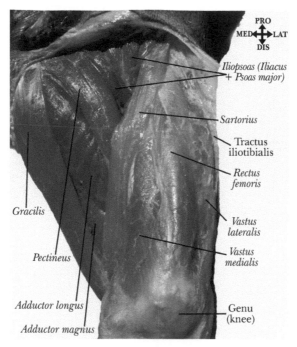

Fig. 59 *Gorilla gorilla* (VU GG1, adult female): ventral view of the muscles of the left thigh.

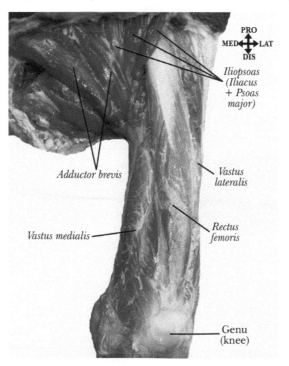

Fig. 60 *Gorilla gorilla* (VU GG1, adult female): ventromedial view of the muscles of the left thigh after removal of sartorius, gracilis, pectineus, adductor longus and adductor magnus.

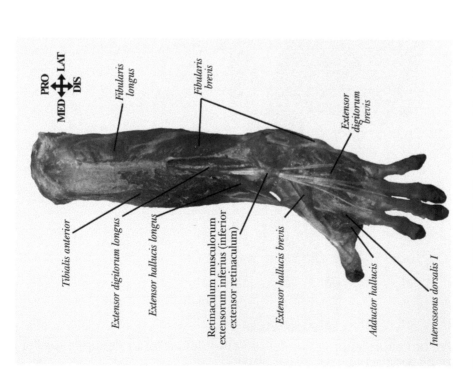

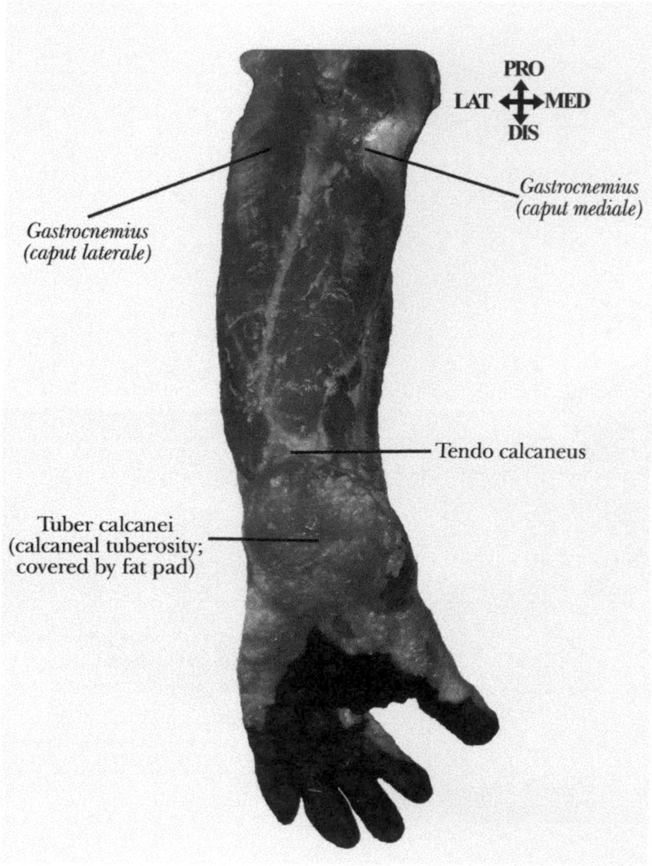

Fig. 62 *Gorilla gorilla* (VU GG1, adult female): dorsal view of the flexor musculature of the left leg.

Fig. 61 *Gorilla gorilla* (VU GG1, adult female): ventral view of the extensor musculature of the left leg and foot.

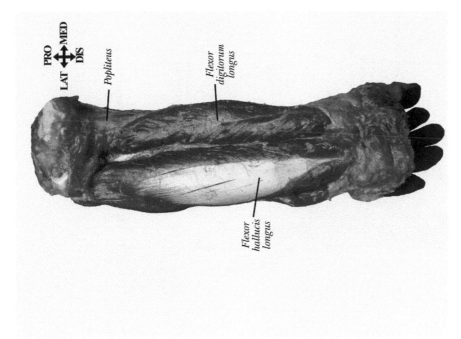

Fig. 64 *Gorilla gorilla* (VU GG1, adult female): dorsal view of the deeper flexor musculature of the left leg.

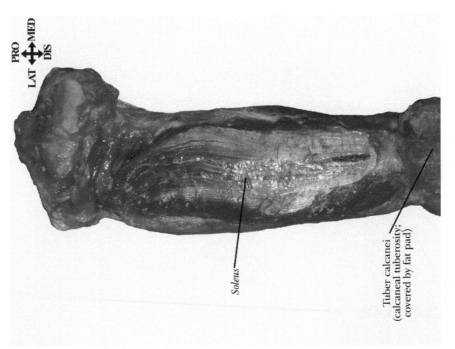

Fig. 63 *Gorilla gorilla* (VU GG1, adult female): dorsal view of the flexor musculature of the left leg after removal of gastrocnemius.

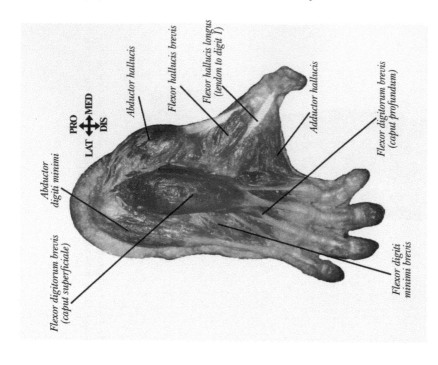

Fig. 66 *Gorilla gorilla* (VU GG1, adult female): plantar view of the muscles of the left foot after removal of aponeurosis plantaris.

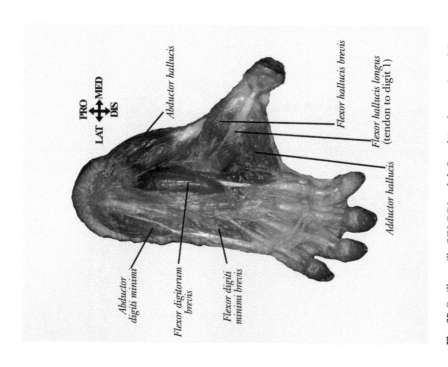

Fig. 65 *Gorilla gorilla* (VU GG1, adult female): plantar view of the muscles of the left foot.

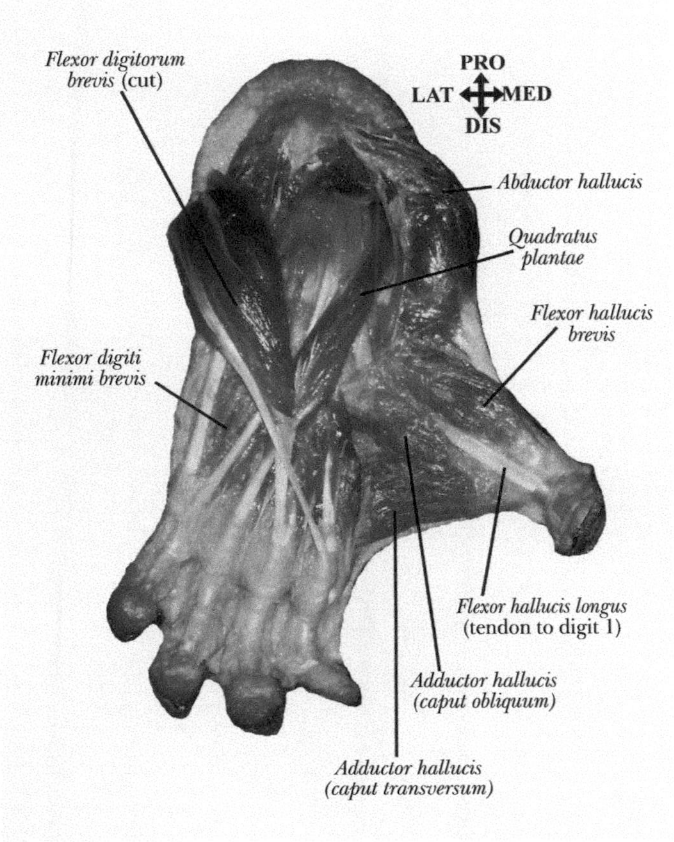

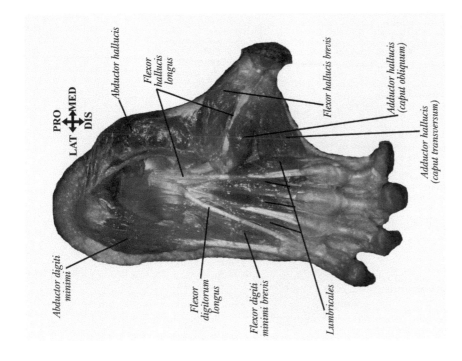

Fig. 68 *Gorilla gorilla* (VU GG1, adult female): same view as in Fig. 67, but flexor digitorum brevis and quadratus plantae were also removed.

Abductor digiti minimi

Flexor digitorum longus

Flexor digiti minimi brevis

Lambricales

PRO
LAT ← → MED
DIS

Abductor hallucis

Flexor hallucis longus

Flexor hallucis brevis

Adductor hallucis
(caput obliquum)

Adductor hallucis
(caput transversum)

Fig. 67 *Gorilla gorilla* (VU GG1, adult female): same view as in Fig. 66, but flexor digitorum brevis was also cut.

Flexor digitorum brevis (cut)

Flexor digiti minimi brevis

PRO
LAT ← → MED
DIS

Abductor hallucis

Quadratus plantae

Flexor hallucis brevis

Flexor hallucis longus
(tendon to digit 1)

Adductor hallucis
(caput obliquum)

Adductor hallucis
(caput transversum)

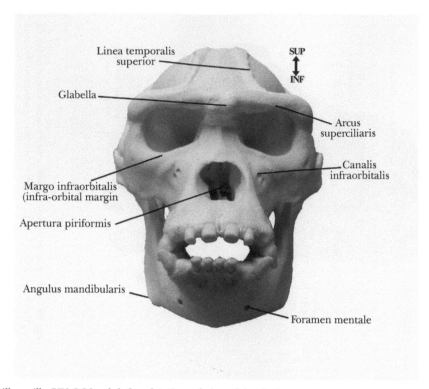

Fig. 69 *Gorilla gorilla* (VU GG2, adult female): Frontal view of the skull.

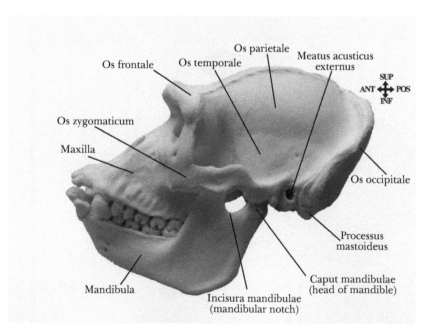

Fig. 70 *Gorilla gorilla* (VU GG2, adult female): lateral view of the left side of the skull.

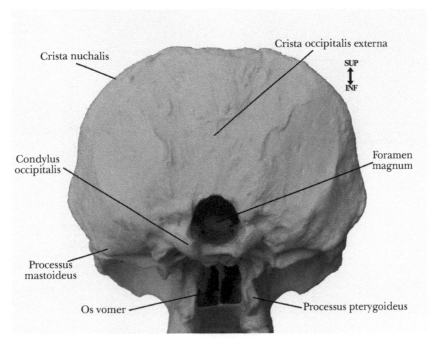

Fig. 71 *Gorilla gorilla* (VU GG2, adult female): posterior (back) view of the cranium.

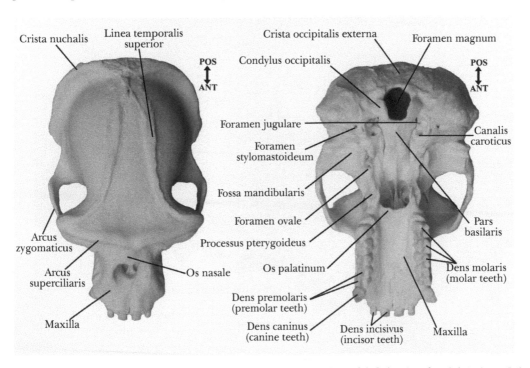

Fig. 72 *Gorilla gorilla* (VU GG2, adult female): superior (on the left) and inferior (on the right) view of the cranium.

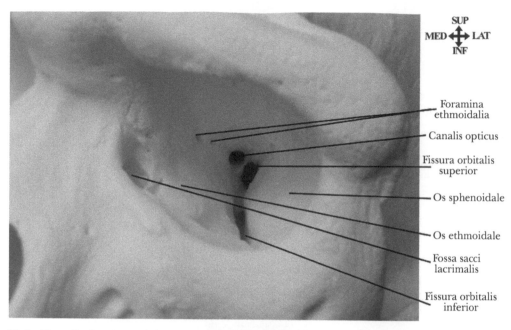

Fig. 73 *Gorilla gorilla* (VU GG2, adult female): frontal view of the left orbital cavity.

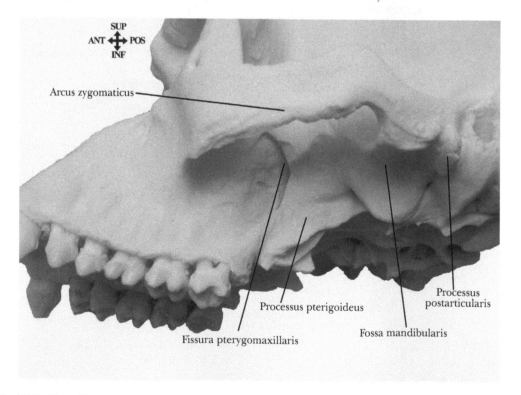

Fig. 74 *Gorilla gorilla* (VU GG2, adult female): lateral view of the left infratemporal region.

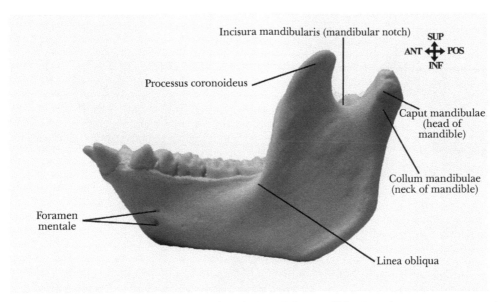

Incisura mandibularis (mandibular notch)

SUP
ANT ◆ POS
INF

Processus coronoideus

Caput mandibulae
(head of
mandible)

Collum mandibulae
(neck of mandible)

Foramen
mentale

Linea obliqua

Fig. 75 *Gorilla gorilla* (VU GG2, adult female): lateral view of the mandible. Note the two separate openings forming the foramen mentale.

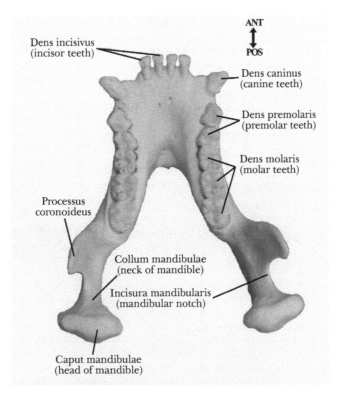

ANT
↑
POS

Dens incisivus
(incisor teeth)

Dens caninus
(canine teeth)

Dens premolaris
(premolar teeth)

Dens molaris
(molar teeth)

Processus
coronoideus

Collum mandibulae
(neck of mandible)

Incisura mandibularis
(mandibular notch)

Caput mandibulae
(head of mandible)

Fig. 76 *Gorilla gorilla* (VU GG2, adult female): superor view of the mandible.

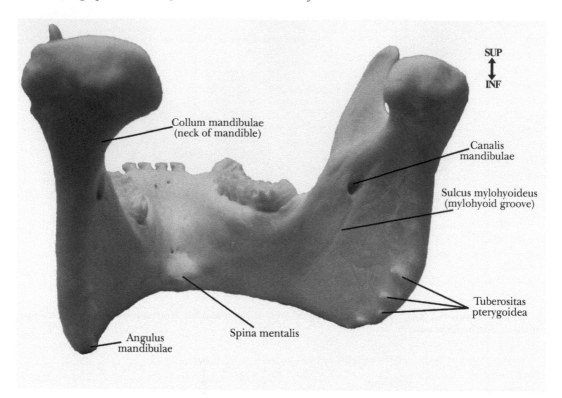

Fig. 77 *Gorilla gorilla* (VU GG2, adult female): posteromedial view of the mandible.

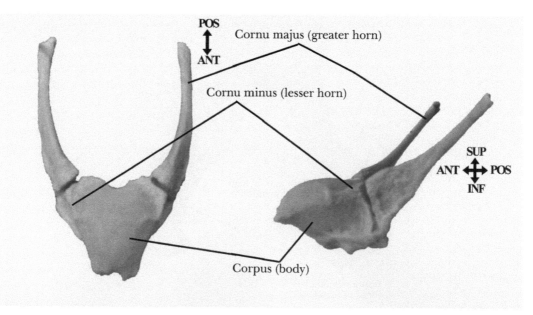

Fig. 78 *Gorilla gorilla* (VU GG2, adult female): inferior (on the left) and lateral (on the right) views of the hyoid bone.

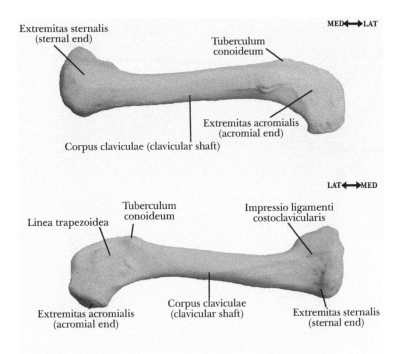

Fig. 79 *Gorilla gorilla* (VU GG2, adult female): anterior (top) and posterior (bottom) views of the left clavicle.

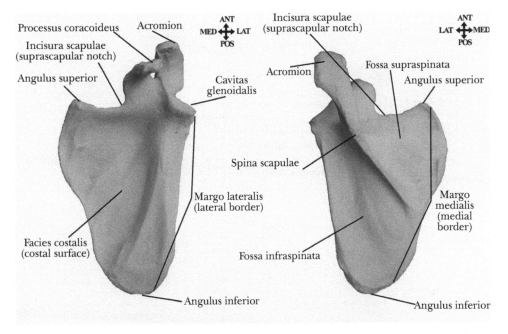

Fig. 80 *Gorilla gorilla* (VU GG2, adult female): ventral (on the left) and dorsal (on the right) views of the left scapula.

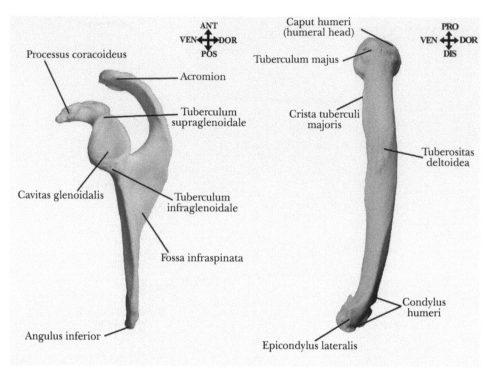

Fig. 81 *Gorilla gorilla* (VU GG2, adult female): lateral views of the left scapula (on the left) and humerus (on the right).

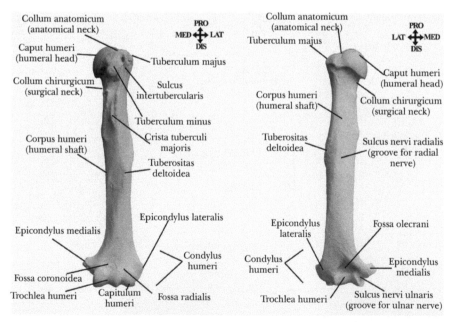

Fig. 82 *Gorilla gorilla* (VU GG2, adult female): ventral (on the left) and dorsal (on the right) views of the left humerus.

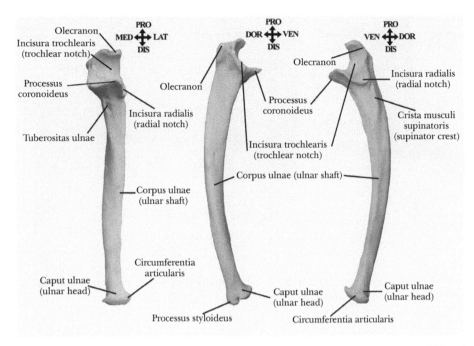

Fig. 83 *Gorilla gorilla* (VU GG2, adult female): ventral (on the left), medial (on the middle) and lateral (on the right) views of the left ulna.

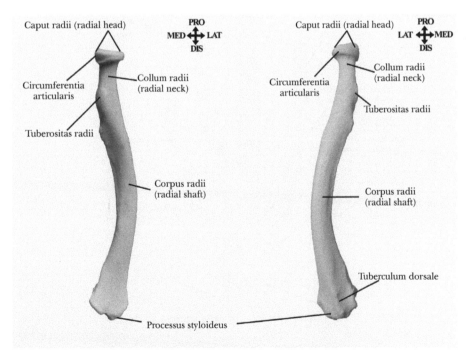

Fig. 84 *Gorilla gorilla* (VU GG2, adult female): ventral (on the left) and dorsal (on the right) views of the left radius.

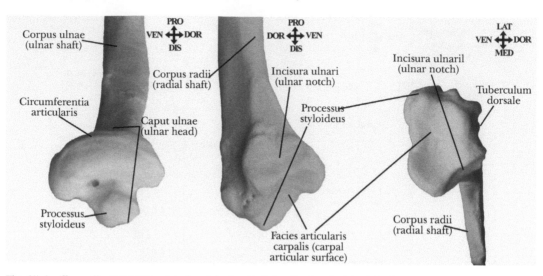

Fig. 85 *Gorilla gorilla* (VU GG2, adult female): detail of the distal end of the left ulna (on the left) and of the left radius (on the middle and on the right).

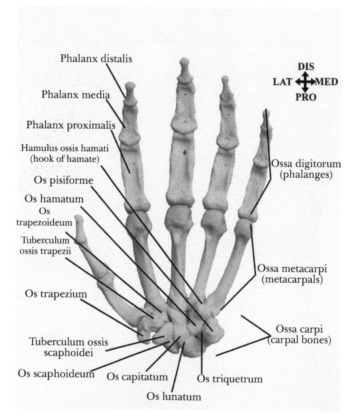

Fig. 86 *Gorilla gorilla* (VU GG2, adult female): ventral (palmar) view of the left hand.

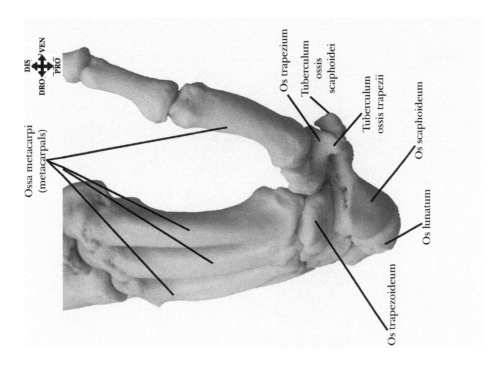

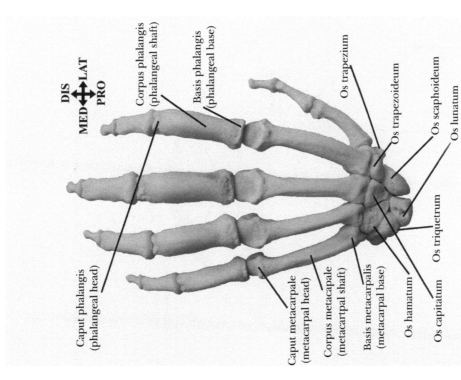

Fig. 87 *Gorilla gorilla* (VU GG2, adult female): dorsal (on the left) and radial (on the right) views of the left hand.

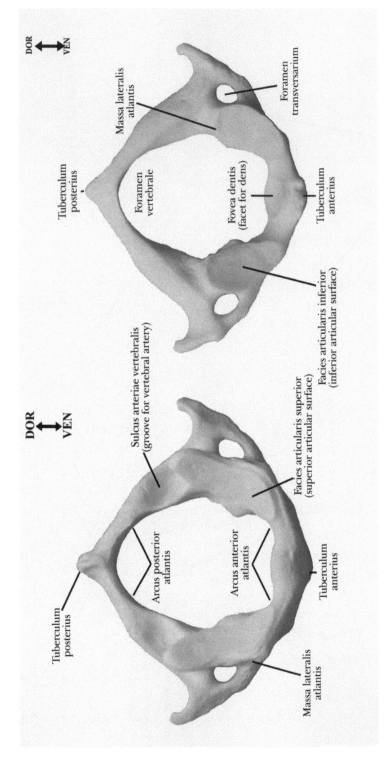

Fig. 88 *Gorilla gorilla* (VU GG2, adult female): anterior (on the left) and posterior (on the right) views of the atlas.

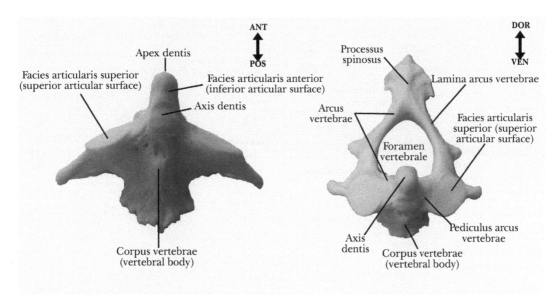

Fig. 89 *Gorilla gorilla* (VU GG2, adult female): ventral (on the left) and anterior (on the right) views of the axis.

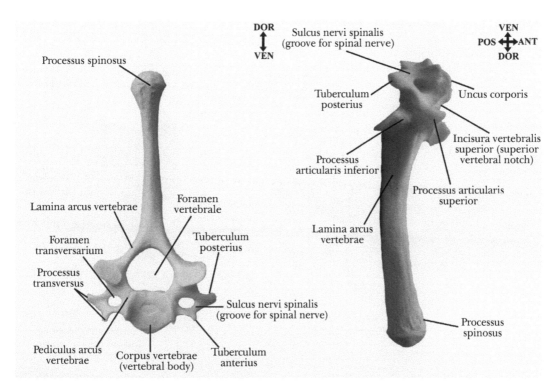

Fig. 90 *Gorilla gorilla* (VU GG2, adult female): anterior (on the left) and lateral (on the right) views of a typical cervical vertebra.

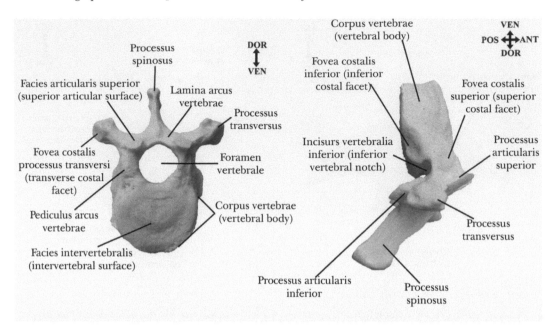

Fig. 91 *Gorilla gorilla* (VU GG2, adult female): anterior (on the left) and lateral (on the right) views of a thoracic vertebra.

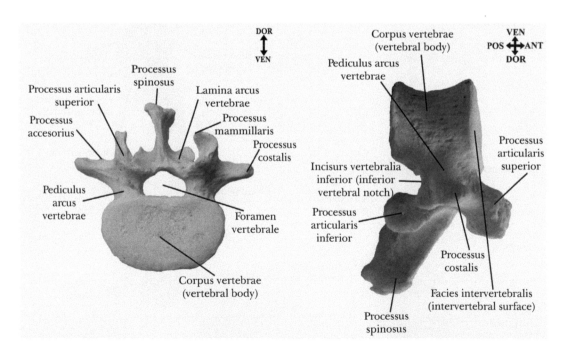

Fig. 92 *Gorilla gorilla* (VU GG2, adult female): anterior (on the left) and lateral (on the right) views of a lumbar vertebra.

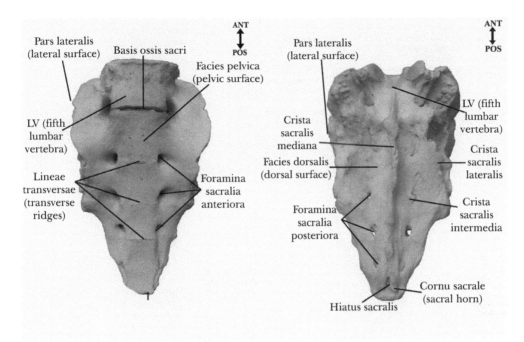

Fig. 93 *Gorilla gorilla* (VU GG2, adult female): ventral (on the left) and dorsal (on the right) views of the sacrum and the fifth lumbar vertebra, which is fused with the sacrum.

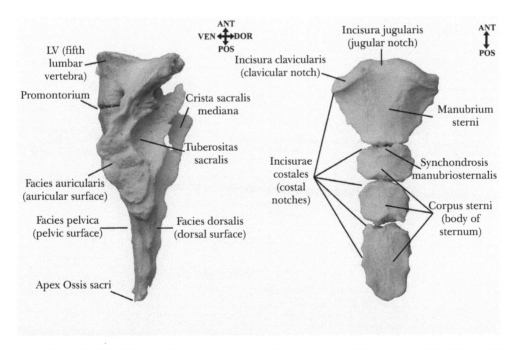

Fig. 94 *Gorilla gorilla* (VU GG2, adult female): lateral view of the sacrum and fifth vertebra, which is fused with the sacrum (on the left), and ventral view of the sternum (on the right; the xyphoid process was not ossified).

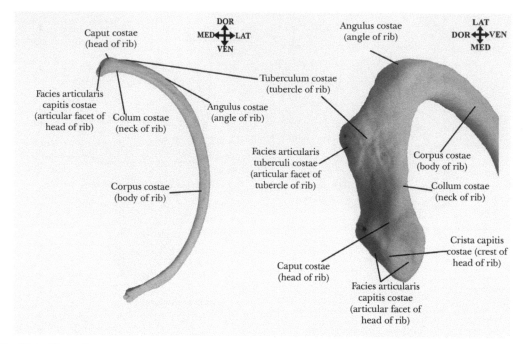

Fig. 95 *Gorilla gorilla* (VU GG2, adult female): anterior view of the seventh left rib (on the left), and detail of the head of this rib (on the right).

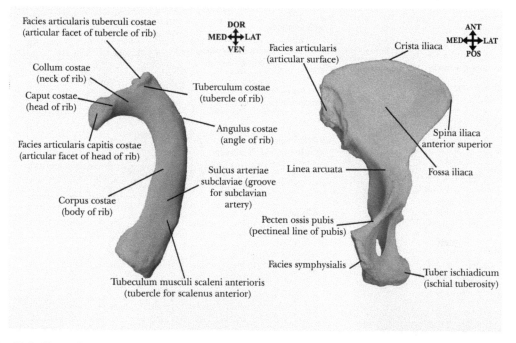

Fig. 96 *Gorilla gorilla* (VU GG2, adult female): anterior view of the left first rib (on the left) and ventral view of the left pelvic bone (on the right).

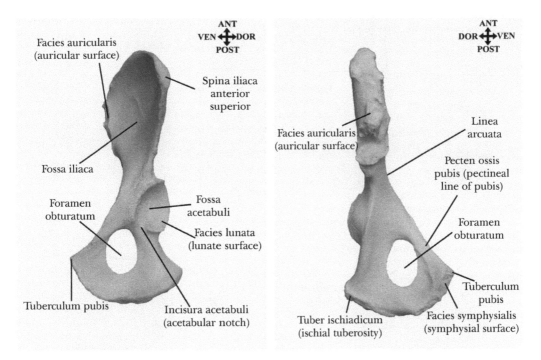

Fig. 97 *Gorilla gorilla* (VU GG2, adult female): ventrolateral (on the left) and medial (on the right) views of the left pelvic bone.

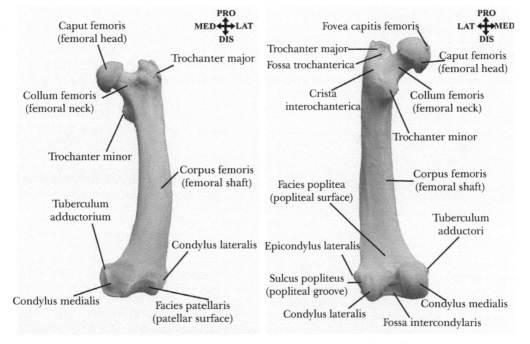

Fig. 98 *Gorilla gorilla* (VU GG2, adult female): ventral (on the left) and dorsal (on the right) views of the left femur.

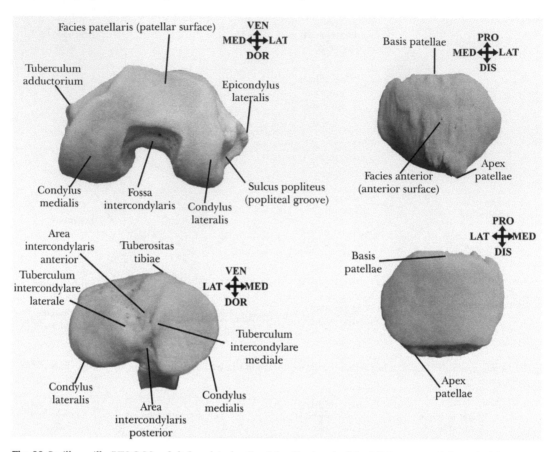

Fig. 99 *Gorilla gorilla* (VU GG2, adult female): details of the distal end of the left femur (top left) and of the proximal end of the left tibia (bottom left), and ventral (top right) and dorsal (bottom right) views of the left patella.

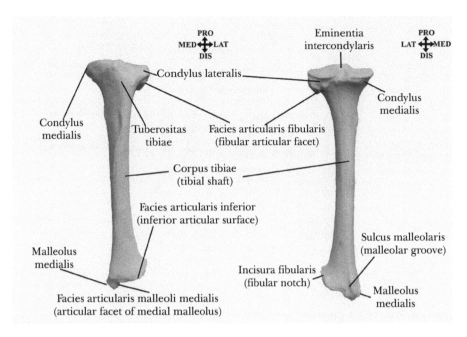

Fig. 100 *Gorilla gorilla* (VU GG2, adult female): ventral (on the left) and dorsal (on the right) views of the left tibia.

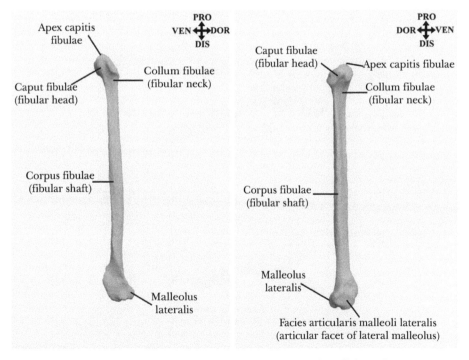

Fig. 101 *Gorilla gorilla* (VU GG2 , adult female): lateral (on the left) and medial (on the right) views of the left fibula.

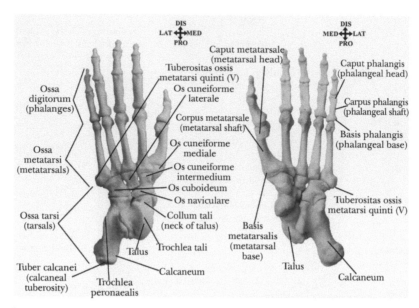

Fig. 102 *Gorilla gorilla* (VU GG2, adult female): dorsal (on the left) and plantar (on the right) views of the left foot.

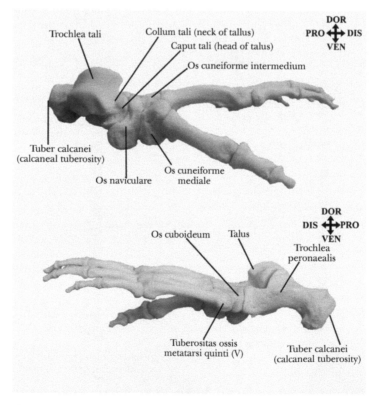

Fig. 103 *Gorilla gorilla* (VU GG2, adult female): medial (on top) and lateral (bottom) view of the left foot.

Milton Keynes UK
Ingram Content Group UK Ltd.
UKHW052027141024
449569UK00016B/733